D0204041

From Caveman to Chemist

From Caveman to Chemist

Circumstances and Achievements

by Hugh W. Salzberg

American Chemical Society, Washington, DC 1991

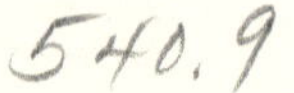

Library of Congress Cataloging-in-Publication Data

Salzberg, Hugh W.
From caveman to chemist: circumstances and achievements
Hugh W. Salzberg.
p. cm.
Includes bibliographical references and index.

ISBN 0–8412–1786–6 (cloth). — ISBN 0–8412–1787–4 (paper)

1. Chemistry—History.
I. Title.

QD11.S23 1990
540'.9—dc20 90–44612
CIP

The paper used in this publication meets the minimum requirements of American National Standard for Information Sciences—Permanence of Paper for Printed Library Materials, ANSI Z39.48–1984. ∞

PRINTED IN THE UNITED STATES OF AMERICA

Second printing 1992

1991 ACS Books Advisory Board

To Emily Mikszto Salzberg
for putting up with a husband
living in several different centuries
at the same time.

Contents

About the Author

HUGH W. SALZBERG spent some thirty-five years teaching chemistry at the City University of New York. Before that, he was in the military service, worked for a time at the U.S. Naval Research Laboratory, and did industrial research. A physical chemist by training, his field of research was electro-organic chemistry. For the past twenty years he has combined his profession, chemistry, with his lifelong love of history and conducted classes in the history of chemistry. Although he is officially retired, he continues his laboratory investigations for the pleasure of it.

Preface

THIS BOOK TRACES the history of chemistry from prehistoric times to the start of the twentieth century. It evolved from lecture notes intended to interest undergraduates in learning something of the origin and development of chemistry and of how chemists arrived at their characteristic viewpoint. I have attempted to supply answers to questions that I wondered about during fifty years of studying, teaching, and practicing chemistry. This book is not a detailed, comprehensive history. It is a narrative designed to give chemists and interested bystanders some insight into the profession.

The story of chemistry is fascinating, full of paradoxes and oddities, false starts and misdirections. Its most striking anomaly is the glaring contrast between the great age of chemical technology and the extreme youth of the science. The technological information that ultimately became chemistry has been collected since the dawn of time. Even before the emergence of *Homo sapiens*, Neanderthals used fire to keep wild beasts from the caves. Later on, people painted cave walls, made pottery, and produced wines and beer. Eventually they used chemical reactions to make glass, metals, and all the other materials needed for complex civilizations. But they made them by trial and error. There was no organized attempt to understand and explain chemical technology until about twenty-five hundred years ago, when the Greeks began to develop science. Unfortunately, the accepted Greek concept of matter led thinkers away from chemistry as we know it now. It was only four hundred years ago, when Greek science lost its authority, that chemistry came into existence.

Chemistry, according to the dictionary, is the study of the composition of matter and its transformations. The term composition refers to both the internal structure of atoms and the way they are arranged within groups. The transformations are the production of metal from rock, plastics from gases, and all the other changes involved in producing the myriad materials our civilization needs.

People have always been fascinated by chemical changes, by changes in color, texture, and other properties when substances react with each other. But not until 1550–1600 A.D. did they display any interest in the internal composition of materials, or even realize that there was an internal composition. Only then did the study of chemistry begin in earnest, but it still took another two hundred years for the science of chemistry to come into its own. Scientific chemistry is actually younger than the United States. Around the year 1770, both Napoleon Bonaparte and Thomas Jefferson believed that matter consisted of earth, air, fire, and water. Yet only thirty-five years later, by the time Napoleon had crowned himself emperor and was selling Louisiana to Jefferson, chemists had worked out the first atomic and molecular weights.

A peculiarity of the story of chemistry is the connection between chemistry and alchemy, that mysterious pseudoscience. Much of the apparatus of early chemistry, and many of the processes, were the same as those of alchemy. The terms and the names were often the same. In fact, many early chemists were also alchemists. Yet chemistry and alchemy were very different, and chemistry did not arise from alchemy.

In the following pages I have traced the development of the basic ideas of classical chemistry, from early times to the end of the nineteenth century. I have focused mainly on evolving concepts of the internal composition of matter, its invisible submicroscopic patterns. Concentrating on the ideas and accomplishments of prechemists and chemists, I intend to show what they thought, what they actually saw and did, and what they thought they were seeing and doing. To understand these people, we must know something of the societies in which they lived, their institutions, the economic possibilities, and the political events that shaped them and their ideas.

Whenever possible, I have discussed the great scientists whose efforts resulted in the intellectual frameworks of chemistry. I hope the reader will feel, as I do, an admiration for them and their astonishing achievements.

I have cited specific references only when it seemed necessary to support a statement. The citations usually refer to standard materials available at most libraries. However, I have also provided a list of background reading for those whose appetites have been whetted.

Hugh W. Salzberg
250 West 94th Street
New York, NY 10025

May 23, 1991

Acknowledgments

To Emily Mikszto Salzberg for help in copy editing and for informed criticism.

To Professor Madelaine Cosman, Director of the City College Institute for Medieval and Renaissance Studies, for volunteering me into giving the lectures that eventually became this book.

To Professors Kaikhosru Irani and Martin Tamny of the City College Philosophy Department and Daniel Greenberger of the City College Physics Department for encouragement, information, and advice. All three are members of the City College Institute for the History and Philosophy of Science and Technology.

To Professor T. Leon Blaszczyk, now of Warsaw University, Poland, for greatly appreciated help in avoiding historical solecisms.

To the librarians from whom I have received so much help: Estelle Davis, Helga Moody, Steve Janofsky, Anabelle Meister, Simone Goldman, and all the others.

To my students who were exposed to the who, the how, and the why in addition to having to learn to use the equations.

To Dr. Robert Multhauf, for his gentle criticism, and to my colleagues Ted Brown, Amos Turk, Sam Wilen, Ron Birke, the late Alois Schmidt, and the late Seymour Mann, for their sympathy and encouragement.

To ACS Books Department editors Robin Giroux, Donna Lucas, and Beth Pratt-Dewey for their help.

I

Ancient Technology

The Roots of Chemistry

THE SCIENCE OF MODERN CHEMISTRY came into existence only in the last half of the sixteenth century, but its origins go back to the Stone Age. Early humans just picked up and used naturally occurring materials. Later, useful substances were made by simple procedures like crushing, burning, and boiling. Still later, complicated processes such as those involved in producing glass were developed. When the Greeks first began to organize empirical observations into science, between 600 and 500 B.C., there was already a considerable body of chemical information. The early Egyptians and Mesopotamians possessed much scientific knowledge. The Egyptians were especially informed about medicine and anatomy, and the Mesopotamians about mathematics, including elementary algebra and astronomy. However, they never organized their information into a philosophic system independent of temple control. As far as we know, before the Greeks, people had no concept of laws of nature independent of the gods' whims.

The Early Stages

The earliest processes that we can call chemical were either one-step reactions that were discovered almost inadvertently, or else were developed step by step, over a long period of time, with each stage resulting in something useful.

Pigments

Since the Stone Age, people have used pigments to color their houses, their boats, their statues, their pictures, and themselves. A sense of aesthetics has always been a factor in humanity's cultural development, and from the beginning, people were attracted by strikingly colored minerals, such as the green malachite, the light blue turquoise, and the dark blue lapis lazuli. They played with them, hammered them, and crushed them—and inevitably noticed that handling these rocks colored their skin. From accidental coloring to deliberate coloring was but a short step. Between 30,000 and 20,000 B.C., people began to grind these beautiful minerals into powders and to use them as pigments, either dry or dispersed in water or oil.

Pigments were used for cave paintings, such as those in France, Spain, and North Africa. They were also used to color the bodies of participants in religious and magical rites, as they still are in New Guinea. And pigments were used as cosmetics. People painted their faces and especially their eyelids. In early Egypt, the upper eyelid was painted black with powdered galena (lead sulfide) and the lower lid green with powdered malachite. In Mesopotamia eyelids were painted with stibnite (antimony sulfide), which is red in some forms and yellow in others. The original purpose may have been to protect against fly bites, but inevitably the emphasis shifted from repelling insects to attracting other human beings (*1*). In Egypt, lips and cheeks were painted with red ochre, while in Mesopotamia, people preferred yellow ochre. Women painted their nails red and tinted their palms and soles with henna. In East Asia, naturally red palms and soles are still considered a sign of good luck.

Dyes

From the earliest historic times, fabrics were dyed. Prehistoric peoples may have used dyes, but cloth is so perishable that we have no sample of dyed fabric that can definitely be traced to the Stone Age. However, prehistoric rock paintings in the Sahara do seem to show people wearing colored garments, and ancient wall paintings in Anatolia (in modern Turkey) show what looks like dyed woven carpets.

Dyes are complicated organic chemicals, usually of vegetable origin, that are generally applied by steeping the cloth or fabric in a dye solution. Perhaps the idea of dyeing first came when people noticed the stains left by walnuts or berries or some other foods. Or the first dyes may have resulted from accidentally wetting cloth with spilled soup or stew made by boiling berries or other plants in a clay pot. Once the first fabrics had been colored by an organic material, it would then have been natural to see what would happen when other plants or seeds were crushed or boiled.

Blue dyes were extracted from woad and indigo; yellow from pomegranate, safflower, and saffron; and red from madder and henna. Kermes, a red dye, was made from animal material: A species of insect growing on the kermes oak was collected, dried, and crushed, and the coloring material was then extracted, probably with an oil. The Phoenicians[1], who inhabited a few towns on the Mediterranean coast of what is now Lebanon, used a local species of mollusk, *Murex brandaris*, to prepare purple dye. They extracted the active secretion from the shellfish in a complicated process, and then steeped garments or cloth in a boiling solution of the extract. Only a minute amount of material could be obtained from each mollusk, so hundreds of them had to be killed for each gram of dye (2). Tyre, the center for the production of purple, had a well-deserved reputation for its foul odor, which emanated from huge quantities of rotting shellfish. The extract was so expensive that only the very rich could afford purple, and ultimately the royal purple became a symbol of nobility. In the Byzantine Empire, an emperor who was the son of an emperor was called *porphyriogenatos*, meaning "born in the purple."

Except for indigo and Tyrian purple, the ancient dyes were not colorfast; they tended to run and to wash out unless certain substances, called mordants, were used to fix the dye to the cloth. Alum (potassium aluminum sulfate) was the most important of these. It was used as early as 1000 B.C. and had international economic importance during medieval times.

But people wanted to color more than their fabrics. The Phoenicians made hair dye from alum, oil of cedar, and the sap of the shrub *Athemis tinctoria*. The resulting hair color was probably light blond or red and must have attracted attention in a region where the population was overwhelmingly brunette.

---

[1]Close to Egypt and Syria and at the western end of the trade routes from Mesopotamia and the Far East, the Phoenicians were ideally located for the trade that was essential to their existence. They picked up all sorts of cultural information and passed it along. Their most important contribution was the Phoenician alphabet, a phonetic script they passed on to the Jews and to the Greeks. As modified by the Greeks, it is the basis of our own alphabet, and a modernized version is still used by the Jews.

Ointments and Perfumes

Oils, especially perfumed oils, were used to protect skin from the Near Eastern sun. Perfumes were extracted from flower petals and fruits by spreading them on layers of purified animal fat, which absorbed the fat-soluble oils and fragrances. Perfumers still use a similar process. The perfumed fat was then shaped into balls and placed on the heads of participants at feasts and allowed to melt and run down. (Psalms 133:2, "The precious ointment upon the head that ran down upon the beard, even Aaron's beard.") Perfumers also developed other techniques for obtaining perfume from fragrant materials. They squeezed aromatic oils out of leaves and fruits by pressing them in a cloth bag. An early kind of perfume-distilling process was done by or under the supervision of women, perhaps because many early chemical procedures and equipment evolved from cooking. Cuneiform tablets, dating from the reign of the Assyrian king Tukulti Ninurtu I (1256–1209 B.C.), mention the name of "the perfumeress" and her female assistant (3).

Fermented Drinks

Because fruits, juices, milk, and other organic liquids ferment naturally, a variety of fermented drinks and dairy products was available very early. People made wine and beer from the very start of historic times—and probably long before. Wine was the favorite drink everywhere in the Middle East except Babylon, where beer was preferred. The Bible repeatedly mentions vineyards, wine, and drunkenness. Noah was found by his sons in a drunken stupor. Lot's daughters (Genesis 19:30–36) got him drunk in order to seduce him into impregnating them. Hannah, mother of the Prophet Samuel, was mistakenly admonished by the high priest to "put away thy wine from thee." Nebuchadnezzar II, the mighty Chaldean king, imported wines on a large scale. Sargon II of Assyria had special wine cellars built into his palace. Ashurbanipal (668–626 B.C.), the last great Assyrian emperor, was something of an expert on wines. He compiled a list of what he considered the best wines, probably the world's first "top ten" list. In Egypt the tax inspector assessed the quality of the wines. In both Egypt and Mesopotamia[2] wines were put into jars at the place of origin in the presence of government officials

[2]The history of ancient Mesopotamia spans some 3500 years, more than half of all recorded history, and it is very difficult to determine which of the many Mesopotamian cultures and political entities was connected with a particular development. Therefore, unless a definite attribution can be made, I use the general terms *Mesopotamia* and *Mesopotamian.*

A wall painting of an Egyptian feast. Each lady has on her head a cone of perfumed fat or tallow. (Reproduced with permission from the Trustees of the British Museum.)

and sealed with an official seal, the equivalent of today's estate-bottled label.

In Mesopotamia, brewing was a home industry, and women were the brewers and beer vendors. (In the Sumerian period, ending about 2600 B.C., women had a relatively privileged position in society. They owned property, had dower and divorce rights, could make legal agreements and sign contracts, and could follow their own choice of occupation. In both Syria and Mesopotamia there was a tradition of women working as distillers, chemists, and alchemists, persisting until at least the Hellenistic era, 320 B.C.– c. 80 B.C.) In the code of King Hammurabi, the Babylonian lawgiver, the women beer vendors were warned not to sell beer of too low a strength nor at too high a price and not to allow political conspiracies to be hatched on their premises. There is a decidedly modern ring to the last restriction. The police watched suspiciously as drinkers chatted—just as Russian and Austrian police did in the nineteenth and twentieth centuries when conspirators met in wine gardens and coffee houses, using social occasions as cover.

Basic Materials from Chemical Processes

Dyes, pigments, perfumes, cosmetics, wines, and beer can all be fairly easily obtained from readily available materials. Other important substances are not so easily obtained. Soap, metals, glass, and plaster are the end products of complicated processes in which the final product is very different from the starting material.

Soap

By the third millennium B.C., the Sumerians were using solutions of soap, although solid soaps were not made until early medieval times (4). Soap was produced—and still is produced today—by rendering, that is, by boiling plant and animal fats and oils with alkaline solutions. Today we make these solutions from sodium or potassium hydroxide. In ancient times, however, they were prepared by leaching the ashes of wood fires with water to form a dilute solution of potash (potassium carbonate) and soda (sodium carbonate). This solution was boiled, and fat or oil was dropped in. The fat or grease dissolved slowly and formed a dilute solution of soap, which could then be used for cleansing or removing grease. Such a complicated process could not have been devised in one fell swoop. There must have been a series of steps or procedures that evolved slowly, each step resulting in a product or process useful enough to be adopted in its own right. Then the new product or process served as the basis upon which the next step was built.

One possible sequence is as follows. Stone Age people used sand or ashes to scrape grease off their hands. If they rinsed away the ashes with water, they might have noted that the water felt slippery. (The ash–water mixture felt slippery because it contained alkali that dissolved the outermost layers of skin.) At a later stage in history, the Sumerians used a slurry of ashes and water to remove grease from raw wool and from cloth so that it could be dyed. (Most dyes do not adhere to greasy cloth.) Sumerian priests and temple attendants purified themselves before sacred rites, and in the absence of soap, they too probably used ashes and water. At some point, people noticed that the more ashes there were in the water (that is, the more concentrated was the alkali solution), the more slippery the water was and the better it cleaned. In fact, the slippery water cleaned even after the ashes were no longer present, and eventually someone put two and two together and discarded the wet ashes.

The slippery solutions clean because the alkali reacts with some of the grease on an object and converts it into soap. The soap then dissolves the rest of the dirt and grease. The more grease and oil dissolved by the alkaline solution, the more soap there is and the better the mixture cleans. People would inevitably notice this because they used the slippery solutions repeatedly until the solutions lost their potency. Thus the Sumerians, realizing that a little grease improved the performance of the alkali, proceeded to make soap solutions directly by boiling fats and oils in the alkali before using it for cleaning. Specific directions for making different kinds of soap solution have been found on cuneiform tablets. The final, almost trivial, step to solid soap was taken about 800 A.D., in Gaul, perhaps in the town of Savona, when soap was separated from water by salting out, or adding salt to, the solution. (Science historian Martin Levey, however, reports that the Mesopotamians used salting out by 3000 B.C. [5].)

Pottery

Far more important than soap were the various products of pyrotechnology: metals, glass, lime, ceramics, and plaster. According to Greek legend, the Titan Prometheus stole the secret of fire and gave it to humanity. Obviously, the Greeks realized that civilization sprang from fire. Without energy from the combustion of wood, we could not have produced the pottery, bricks, lime, weapons, and tools that enabled us to move out of the Stone Age.

Pottery was the first product of pyrotechnology. Perhaps the sequence of discovery went like this. At first, primitive humans placed cooking fires on the ground, but eventually they used cooking pits, holes in the ground lined with firewood. If the pit had been dug in mud or clay, the walls of the pit slowly turned to stone, or rather, to brick. Clay can be thought of as water-softened stone, or stone containing water. When the water is driven out by heat, the clay changes back to hard stone; this change is what takes place in the pottery kiln when ceramics are produced. Clay exposed to the heat of a cooking pit eventually becomes rock-hard. If a lump of clay was dropped into a roaring blaze, either accidentally or deliberately, it too would harden. Once people noticed that the lumps had hardened, they began to shape the soft clay into rudimentary human and animal figures, heated them, and formed ceramic figurines. Archaeologists have found baked clay images that date back twenty thousand years. More practical experimenters plastered gourds and wicker baskets with clay and baked them to produce pots, bowls, and jars. After about 6500 B.C., the potters finally dispensed with gourds and baskets and shaped pots, bowls, and jars directly from wet clay.

Metals

More than pottery came from the fire pits. To get greater heat, the potters tried different sizes and shapes of the fire pit. Perhaps as early as 6000 B.C., they had transformed the fire pit into a furnace that could reach temperatures high enough to melt copper. By 5000 to 4000 B.C., artisans could produce furnace temperatures high enough to transform copper ore into metallic copper. With further trial-and-error improvements, other metals were produced, as well as lime, plaster, and glass.

The great advantage of metals is their plasticity. Metals can be stretched, bent, and melted, and so metal objects can be easily shaped and reshaped. Wood is easier to cut or bend than metal, but it wears out quickly and is fragile. Stone is more durable than wood and harder than either wood or metal, but it is brittle and difficult to shape. A cracked stone arrowhead or a broken wooden spear is difficult or even impossible to repair, but a broken or bent metal blade can be hammered back into shape or even melted and reformed. Moreover, metals have a variety of uses unsuitable for stone and wood.

High civilizations like those of the Egyptians, Mesopotamians, Chinese, and Greeks were dependent on metals. For example, the Egyptians and the Greeks used iron tools for mining and copper saws for cutting stone. The Palestinians and the Greeks farmed with iron plowshares. The Greeks trimmed and smoothed stone with iron chisels. They had metal clasps and pins for clothing, metal hoops for barrels, and metal hinges and other furniture fittings. They used metal clamps, spikes, pins, and dowels for stone columns, and they used structural iron beams in many of their temples. (In Sicily, at Acragas, the modern Agrigento, the temple had iron beams 15 feet long and 5 by

12 inches in cross section.) They had metal hammers, spikes, drills, files, rasps, knives, nails, arms and armor, and screws. The Egyptians and Mesopotamians too used many of these implements as well as metal tires for weapons and chariots.

Ancient civilizations were not, however, dependent on metals in the way that ours is. Metals appeared in quantity during the Neolithic era, and their uses developed gradually, but not until the spread of the use of iron for tools and weapons, about the middle of the first millennium B.C., did people begin to lose their dependence on stone, bone, and wood. Until about 1000 B.C., most Egyptian and Mesopotamian peasants had almost no contact with metals except for the ornaments in the temples and the jewelry, arms, and armor they saw on nobles and high-ranking officials. People used wooden shovels, rakes, wheels, and mill wheels until well after the onset of the Industrial Revolution, especially in underdeveloped areas.

The first metals known were copper and the relatively scarce gold and silver. Under ordinary conditions, most other metals react rapidly with oxygen in the air, so tin, mercury, iron, lead, zinc, and other common metals are found only in combinations with oxygen and/or other nonmetals. To recover metal from these combinations requires sophisticated and fairly drastic chemical treatment. Gold, on the other hand, is very unreactive and is found only uncombined as metal. Silver, too, does not react easily with oxygen and is found uncombined except in formerly volcanic areas, where it has reacted with sulfur. Copper does react with oxygen but slowly. Only gold and silver and perhaps some copper could exist as lumps of native metal lying around on the earth's surface waiting to be noticed.

Gold and copper were first discovered about 9000 to 7000 B.C. in the form of nuggets or lumps that looked like stones. Most likely, Neolithic toolmakers picked up these stones and attempted to form them into arrowheads or knives. To the astonishment of the toolmaker, the stones did not chip, split, or flake the way stones normally did; they bent and changed shape like clay—a totally bewildering behavior. Nothing in Neolithic experience had prepared them for rocks that bent. It was magical. Thus began the long association of metals with magic.

The toolmakers played with the fascinating pebbles, doing different things to them to see what would happen. They found that after they were bent, they could be straightened out again and bent into different shapes. They could be beaten into flat plates and cut into patterns. These new forms were ornamental, thought to be magical, and were in great demand—gold especially, because it never loses its beautiful luster. Whether gold ornaments conferred status on the owner or were appropriated by those who already had status, chiefs and shamans wore gold, silver, and copper jewelry, and the metals rapidly became associated with rank and power.

It is relatively easy to find and use metals; it is more difficult to produce them. Thousands of years passed between the discovery of natural metals and the first smelted metals from ores. For one thing, the relationship of metals to their ores is not at all obvious. Although minerals had long been used for pigments, no one dreamed that there was any connection between the beautiful colored minerals and the shiny metals. Moreover, smelting is a very complex and difficult process.

Copper was the first metal produced from its ore because it is the easiest to smelt. Probably the ore was malachite (hydrated copper carbonate). When malachite is heated with wood or charcoal at a high temperature, the carbon and carbon monoxide from the wood remove the oxygen from the ore and reduce it to metallic copper. (*Reducing* refers to the chemical reactions; *smelting* is the term for the overall process.) The first smelting might possibly have taken place in a cooking fire, but the chances are that cooking temperatures would have been much too low and that oxygen in the atmosphere would have prevented reduction. Smelting requires high temperatures such as those in a pottery kiln, so a kiln is probably where the first metal was produced. Perhaps a greenish copper pigment was used to decorate a clay pot before firing and the potter found a film of copper on the pot afterward. (However, no early pots have been found with copper ornamentation.) Or maybe a lump of malachite was used to prop up a pot. More likely, human nature being what it is, someone deliberately put some rock or powdered mineral into the furnace or kiln out of curiosity, just to see what would happen. Just for the hell of it! In any event, the result must have been astounding. The blue or green rock had disappeared and red copper was present in the furnace. Somehow a rock had turned into copper. The experiment must have been done again and again, and soon other rocks were put into the furnace to see what would happen. As a result, other metals and materials were produced, the metals needing higher smelting temperatures appearing last. Copper, tin, and bronze were produced in economically useful quantities by 4000–3000 B.C., smelted iron by about 1200 B.C., and, finally, zinc in the medieval era, although small amounts of zinc had been made in Roman times. In China, iron ores contain relatively large amounts of phosphorus and melt at lower temperatures. Iron, therefore, appeared in China at a much earlier date than in the Middle East.

Glass

Glass is another product that came out of the pottery furnace. It is the world's first synthetic thermoplastic. Molten glass can be poured into almost any shape and retains that shape upon being cooled. Actually glass is a liquid that, at room temperature, is too viscous to flow except very slowly and under pressure. Like most liquids, it is transparent unless it contains air bubbles or undissolved particles. Ancient glass contained much undissolved material and so was not as transparent as ours.

Glass was first produced as a byproduct of metallurgy. In mining, a good deal of rock is unavoidably dug up along with the ore, and when the ore is heated, the unwanted rock is heated too. The initial steps in smelting convert the ore to metal but leave the rock unchanged. The result is an unusable mixture of rock and particles or lumps of metal. The earliest smiths had to break up the mixture and pick out the metal, piece by piece. As time went on, however, improved furnaces could attain temperatures hot enough to melt both the metal and the rock to form two white-hot liquids that, fortunately, were insoluble in each other, like oil and water. To separate them, all the smith had to do was to pour one of the two liquids into a receptacle. This separation process is called *liquation*.

The early smiths soon found that, after liquation, when the molten rock cooled down, it formed a very interesting new material, a rigid, glassy solid, or slag, very like the prized volcanic material, obsidian. Studying this new material by trial and error, heating all kinds of rocks, and eventually purifying their starting materials before smelting, they produced slags that were blue, green, brown, or even red—and some that were colorless and transparent. Actually they had made glass, and it didn't take long before they were using the glass for beads and other ornaments, figurines, and even flasks and beakers.

Other materials were also put into the furnace to see what could be made from them. At least as early as 4000 B.C., quicklime was produced from limestone by driving off carbon dioxide. Quicklime was, and still is, used to remove fat and hair from animal hides in the manufacture of leather. Since about 300 B.C., however, quicklime's major function has been in making cement for construction. Plaster of Paris was also produced by heating gypsum.

Magic and the Furnace

Whatever the end product, the dramatic changes taking place in the blinding flames and searing heat of the furnace seemed magical. Rocks turned into metal and glass. Obviously this was the work of gods or demons. So, gods and priests of metallurgy arose. Each people had its own god of the forge. He was Hephaistos to the Greeks, Vulcan to the Romans, Welland Smith to the Saxons, and Tyr the One-Armed (from whose name we get Tuesday) to the Norse. Everywhere, rites, sacrifices, and incantations were developed to keep the gods in good humor or to appease the demons.

These rites had practical functions too. They must have been fairly effective in ensuring correct heating times and furnace conditions, which must be carefully controlled. For the smith of ancient times, with no way to measure temperature, tell time, or determine composition of ore, control was almost impossible. Consequently, things often did not work. (In a diplomatic note of the thirteenth century B.C. [6], Hattusilis III, king of the Hittites, apologized to an Assyrian king, "I have no good iron in my warehouse right now . . . " and promised to send some of the next good batch.) When something went wrong, people thought the gods must have been offended, and they offered a propitiating ceremony. On the other hand, when things did go right, it was imperative that the illiterate smith remember exactly what he did so that he could repeat it the next time. The ritual chants and incantations must have been excellent mnemonic devices to help the apprentice learn correct procedures and timing.

From the very beginning, metallurgy and pyrotechnology were connected with magic and ritual, a connection that eventually carried over into esoteric, or religious, alchemy. Much technological information was kept secret, lest it fall into sacrilegious hands. In both Assyria and Egypt, there were written recipes and lists of materials, but as early as 1700 B.C., papyri and cuneiform tablets contained warnings to the reader not to divulge secrets, coupled with curses on those who did (7). The Assyrians, who were enthusiastic antiquarians, got most of their information from ancient Sumerian texts. Some of these were simple and straightforward, but many used a cryptic language intended to be understood only by the initiates. Copper, for example, was sometimes called eagle, and ferrous sulfate, green lion.

The ancients had no concept of natural law, although they clearly understood cause and effect. They lived in a world where

Judgment of a soul after an Egyptian tomb painting. The good deeds are being weighed against the evil deeds in an equal-arm balance. Similar balances are still being used, especially by jewelers and apothecaries.

Nature responded to ghosts, gods, and demons. It rained because the local rain god was either pleased or angry, depending on the circumstances and the locality. Rocks and minerals were alive; they existed in male and female forms; they were born, grew in the ground, died, and had magical powers. To the Sumerians, each metal was associated with a god and a planet; they believed there was a strong connection between chemical and metallurgical processes and the stars, a belief that persisted until at least the seventeenth century A.D. Meteoric iron, which came from the sky, was especially sacred. Until the Greek philosophers' time, technological phenomena were given religious or magical explanations. As far as we know, there was no systematic attempt by the ancients to explain natural phenomena on a philosophical or scientific basis, even though the Babylonians, for example, had accurate astronomical data and could actually predict eclipses.

In ancient Egypt, the temples were centers of learning and intellectual activity, like the monasteries in medieval Europe. Of course, education was not completely confined to the priesthood. Many nobles and officers in the secular bureaucracy were literate, as were physicians, and there were literate slaves to tutor the children of the wealthy. The temples maintained libraries and schools for scribes, masons, jewelers, and all the others needed in the service of the temple gods. Many craftsmen were engaged in the private sector of the economy, working for minor officials and their wives. Imitation gold, silver, lapis lazuli, and other gemstones were always in demand; the cost of real gold, silver, and gems was too great for any but the royal family, the great nobles, and the higher echelons of the priesthood. Always the bulk of Egyptian technology, astronomy, and mathematics was of and for the temples. Naturally, Egyptian chemical processes were permeated with magic and religion.

The situation was similar in Mesopotamia, with some differences due to geography. As in Egypt, the temples were the centers of learning and knowledge. The stars and planets were observed nightly in order to keep track of the calendar of religious events. The god or goddess, through the temple priests, directed the irrigation and flood control works, and to the deity extensive tithes and voluntary offerings were brought. As a result, priests needed to keep accurate records, both of the large quantities of goods brought to the temple and distributed to workers and priests and of the assignments of the work groups and their performance. Writing and mathematics thus began with the temple accounts but were also needed for secular matters. Raw materials such as stone, timber, and metals were scarce in Sumer; its people had to trade to survive. The Mesopotamian economy had a relatively large private sector with many shopkeepers, traders, merchants, and independent artisans. Commercial law was also well developed, and tablets are extant that mention suits for breach of contract. Merchants needed to be literate and able to handle rudimentary arithmetic, and so there were secular scribal schools that trained them and their sons. Such education was mostly practical and limited to matters useful in business, but there was also an extensive written literature including romances, poetry, and drinking and love songs.

Because ancient Egyptian and Mesopotamian centers of learning were temples, the very concept of a nonmagical explanation of eclipses, earthquakes, pyrotechnology, and other phenomena was not only unthinkable but also sacrilegious. A priest who had been trained to believe that certain rites propitiated the god of glassmaking and who also earned his living from the fees and offerings he collected from the glassmakers could not be expected to think about a chemical explanation for the process. If anyone else did offer such an explanation, he would have been considered a threat to both the god and the priest.

Egypt was an isolated and stable society, but perhaps the adaptable Mesopotamians could have freed themselves from the domination of their temples and theocracies and developed the concept of natural law. Unfortunately they did not have the opportunity. After the middle of the second millennium B.C., they were a battered civilization, constantly at war, under attack by foreign invaders, repeatedly under foreign domination, and subjected to sack, pillage, massacre, and wholesale transfer of populations. Under such conditions, they clung to the safety of traditions and did not embark on new philosophical ventures. Fortunately, Mesopotamian and Egyptian knowledge had been transferred to peoples more favorably situated, the Minoans and the Greeks, and from these populations new intellectual developments would come. Natural philosophy had to await the Greeks.

References

1. Forbes, R. J. "Chemical, Culinary and Cosmetic Arts." In *History of Technology*; Singer, C.; Holmyard, E. J.; Hall, A. R., Eds.; Oxford University Press: Oxford, England, 1957; Vol. I, p 292.
2. Pliny. *Natural History*; Vol. IX, p 36, as cited by Levey, M. *J. Chem. Ed.* **1955,** *32*, 627.
3. Ebeling, E. *Parfümrezepte und Kultische Texte aus Assur*; Pontifical Institute: Rome, 1950; pp 32, 46. Cited by Levey, M. In *Great Chemists*; Farber, E., Ed.; Interscience: New York, 1961; p 6.
4. Levey, M. *J. Chem. Ed.* **1954,** *31*, 521–524.
5. Ibid., p 524.
6. Forbes, R. J. *Studies in Ancient Technology*; Brill: Leiden, Netherlands, 1953; Vol. IX, p 254.
7. Forbes, R. J., op. cit., Vol. I, p 125.

Timeline—Ancient Technology

Technological Advances	Mineral pigments; cave paintings	Baked clay figures	Gold ornaments from native gold	Clay pots, bowls	Egyptian eye paints; plaster of Paris; copper, tin, bronze in Egypt and Mesopotamia; Sumerian pictographs	Sumerian cuneiform; soap; Egyptian hieroglyphics	Iron first forged, perhaps from meteoric iron; beer brewed by Sumerians
	30,000 B.C.	18,000 B.C.	9000–7000 B.C.	6500 B.C.	4000–3500 B.C.	3500–3000 B.C.	3000–2500 B.C.
Historical Landmarks					Sumerian city-states in Mesopotamia		Egypt unified in 2800 B.C.

Timeline—Continued

Technological Advances		Hammurabi sets standards for beer	Start of Iron Age in Anatolia, Syria, Palestine	Phoenicians import tin from Britain		Sale of beer regulated in Egypt; Tyrian purple first made; alum used as mordant; iron used in Greece	Steel made in India and Caucasus
	2000–1500 B.C.	1800–1700 B.C.	1500–1000 B.C.	c. 1200 B.C.	1193 B.C.	1000–900 B.C.	900–800 B.C.
Historical Landmarks	Greeks reach Mediterranean	Hammurabi 's law code	Start of Assyrian Empire	Israelites appear in history	Sack of Troy by Greeks		

Timeline—Continued

Technological Advances	Iron household utensils				
	800–700 B.C.	624–545 B.C.	550 B.C.	547 B.C.	490 B.C.
Historical Landmarks		Thales of Miletus lived	Birth of Gautama Buddha	Persian Empire founded	Battle of Marathon

II

Hellenic Chemical Science

Greek science, philosophy, literature, and medicine dominated the Mediterranean world for more than a thousand years—the period from 550 to 320 B.C., known as the Hellenic Era. The Greeks took the technological knowledge of the Egyptians, Mesopotamians, Persians, Syrians, Indians, and perhaps even the Chinese and organized it into natural philosophy, the precursor of both philosophy and science. They originated many of the scientific and philosophic disciplines and posed many of the philosophical questions still being discussed. During the Hellenistic Era, which began with the conquests of Alexander the Great, who died in 323 B.C., the Greeks became the dominant political and cultural element in Syria, Egypt, Asia Minor, Mesopotamia, southern Italy, and Sicily. When Rome conquered that area, starting about 200 B.C., Greece became, geographically, just a minor Roman province, but the Romans both absorbed and imitated Greek culture and eventually transmitted it to the Christian world. Only with the advent of Islam, about 650 A.D., did Greek cultural domination die out in Egypt, Syria, and North Africa. Even then, much of the Greek literature was translated into Arabic and became the basis of Islamic philosophy and science.

Histories of western science usually start with Greece, because there are more source materials available from Greece than from preceding civilizations, and the first western natural philosophers of whom we have knowledge were Greek. Much written material on Greek science is readily available, some in the original language and a good deal in Arabic and Latin translations. In fact, there is so much discussion and commentary in Latin and Arabic that we are often familiar with the contents of Greek manuscripts that no longer exist. Moreover, Greek never became a dead language, like Sumerian. There have always been people who speak and read Greek, not only classicists but also specialists in philosophy, history, science, and mathematics.

The Role of the Alphabet

The Greek alphabet was central to the development of Greek thought. It was most probably adapted from the Phoenician alphabet, with local variations that lasted until the final standardization in 403 B.C. by the Athenian archon Euclides. It was so simple and the phonetic writing system was so easily mastered that almost any Greek could rapidly learn to read and write. Compared with the two dozen letters of the Greek alphabet, there were thousands of Egyptian hieroglyphics, hundreds of Mesopotamian cuneiform symbols, and thousands of Chinese pictographs.

In China, Egypt, and Mesopotamia, literacy was the privilege of those few enrolled in temple schools or whose families could afford to pay for tutors (or who owned educated slaves). In those societies, education was a virtual monopoly of the temple. In the Hellenic world, however, there was no need for temple schools. Professional tutors gave inexpensive, private instruction, and literacy was attainable by virtually all who were interested. As a result, except for those priests officiating at the few shrines, there was no influential organized priesthood. Religion was very important, but it had little economic or political power compared with that in the older societies.

Greek Society and Science

Greek society differed greatly from the societies of Egypt and Babylon. And, therefore, so did the Greek viewpoint—or at least the viewpoint of Greece's educated elite, which is the one we know from its writings. The Greeks lived in small, self-contained units, city-states, where they were for the most part citizens, not subjects or slaves. They owned their own farms, they made their own political and economic decisions, and they took an intense interest in civic matters. They also were interested in science, philosophy, and mathematics. That is, some Greeks were. Not all. Many Greek states contributed nothing to the intellectual discourse. The attitudes that concern us here are primarily those of the educated elite of the cosmopolitan centers, such as Athens, Rhodes, Croton, Syracuse, and Miletus. And even there the educated Greeks were representative of their own times, not ours. Their rationality was mingled with what we would call mysticism. Empedocles, philosopher and statesman (c. 500–430 B.C.), for example, believed that he had supernatural powers. Pythagoras, and perhaps Plato as well, believed in the power of magic numbers.

In science and philosophy, the pioneers were the Ionian Greeks of Asia Minor, especially those of the city of Miletus in what is now Turkey. The Ionian cities were commercial centers, maintaining extensive contacts with Egypt, Mesopotamia, and Persia and trading some goods with countries as distant as China. By about 600 B.C., an influential class of well-traveled, highly cultivated, polylingual Ionian merchants—rational, speculative, and, above all, literate—were the intellectual leaders of Greece. Only after Ionia was subjugated by the Persians about 546 B.C. did Athens come to the fore, and even then the intellectual growth of Athens was stimulated by Ionian refugees.

Hellenic Greek science achieved much, especially in astronomy (prediction of an eclipse), medicine (discovery of the Eustachian tubes), and mathematics (the Pythagorean theorem and discovery of irrational numbers), while operating under some severe handicaps. The entire Hellenic period lasted a relatively short time, some 230 years. Moreover, there were few Greeks, and only a fraction of these were of the privileged male group that had the leisure to study nature. Of the few men available for natural science, most were not interested, and those who were interested were handicapped by their social class's attitude toward arithmetical computation—it was connected with commerce and considered to be somewhat vulgar.

But perhaps the greatest handicap of classical Greek science was that it was largely cut off from an empirical technological base (*1*). The Greek natural philosophers were not really interested in applying their results to the problems faced by the farmer, the smith, the shipwright, and the glassmaker. They had little interest in practical technology and made few contributions to it. They got neither information nor inspiration from daily life, except perhaps from medicine and warfare.

Greek scientists, however, did have some undeniable advantages over earlier investigators. They did not labor for a theocratic state or a temple and could study whatever they found interesting. Also, after the defeat of the Persians, interested Greeks had the opportunity to travel, to compare ideas and cultures, and to think. Travel was relatively cheap and safe because Greek sea power dominated the eastern Mediterranean. Greek gentlemen could and did take the equivalent of the eighteenth century grand tour, and they observed many interesting and unusual phenomena that fed their intellectual appetites.

In most cases, the viewpoint of Greek natural philosophers was surprisingly modern. They were, in general, practical men trying to order the universe in terms of their own common sense and observation. About 400 B.C. the great physician Hippocrates presided over an informal school of physicians on the island of Kos. One member of that school wrote a book on epilepsy, which was known as the sacred disease. Part of the introduction reads:

> I am about to discuss the disease called "sacred". It is not, in my opinion, any more divine than any other disease but has a natural cause and its supposed divine origin is due to men's inexperience and to their wonder at its peculiar character. But if it is to be considered divine because it is wonderful [i.e., not understood] there will not be one but many sacred diseases [2].

Writings like this show a thoroughly rational approach. Of course, there is a danger in this attitude when pushed too far; common sense can sometimes be misleading when applied to natural phenomena.

Early Scientific and Philosophical Concepts

Greek scientists believed in natural law, law that even the gods themselves had to obey. We do not know how and why they arrived at the concept of natural law, but perhaps an important impetus was the sensational achievement of Thales of Miletus (624?–546 B.C.), the first philosopher of whom we have a record. He became famous for predicting the solar eclipse of May 25, 583 B.C. When news of this event spread over the Greek world, thoughtful men realized that eclipses followed patterns and were natural phenomena occurring at definite intervals. Other ancient peoples had believed that eclipses were due to the unpredictable whims or angers of the gods and regarded them with fear and trembling, but educated Greeks knew that even heavenly bodies moved along predictable paths that could be followed mathematically. The Mesopotamian astronomers, from whom Thales had probably obtained his knowledge of astronomical records, had also known this, but since they either were priests or worked for temples, they had every incentive not to publicize their conclusions. We do not know the ideas of Thales's Greek predecessors, if any, but we do know that after Thales, educated Greeks were convinced that nature was rational. While other peoples of the ancient world believed that reciting a magical formula could change the course of nature, the Greek philosophers were trying to explain natural phenomena in terms of natural laws.

Greek philosophers believed that all things had both cause and effect. Pythagoras is supposed to have declared that the world must be rational because God is a geometer. Of course, to Pythagoras rationality was mathematical. He believed that all objects in nature were expressed in terms of integral numbers (which to Pythagoras were physical entities that corresponded somewhat to our atoms) and that all numbers were rational; that is, they could be expressed as a ratio of integers.

Ironically, the Pythagoreans themselves discovered that certain numbers, such as the square root of 2, are irrational; they cannot be expressed as a ratio of integers. The psychological blow was devastating to them. The world of numbers did not correspond completely to the physical world. The consequences of this discovery were of immense importance. Because the Pythagoreans could not accept and work with such irrational numbers, they and the other classical Greeks turned toward geometry rather than numerical computation, at least partly because irrational numbers could be expressed in terms of geometrical constructions, bypassing any awkward philosophical implications.

Most Greek philosophers after Thales seem to have agreed that all material things were composed of a few basic elements or principles, which in turn were formed from a prime matter. They disagreed about how many elements there were and what they were and if they were in the form of atoms or a fluid or a vague spiritual essence. Thales himself believed that all things originally came from, or were composed of, water.

In general, philosophers accepted the doctrine of contrarieties, widespread in the ancient world. To explain cyclical changes such as the rhythm of the seasons and the day following the night, societies ranging from primitive tribes to highly sophisticated nation-states reasoned that all things are a fusion of contradictory and opposing principles, such as dark and light or male and female. The Chinese believed in two such principles, yin and yang, one dark and female, the other light and male;

one passive and the other active. The Persians believed in Ahura Mazda (Ormazd) and Ahriman, one light and good, the other dark and evil; one the principle of daylight and the other of darkness.

The Greek idea of the composition of matter was a somewhat more complex version of contrarieties. First formulated by Empedocles, the idea was used by Plato (c. 427–347 B.C.) in his dialogue *Timaeus*, and then incorporated by the great Aristotle (384–322 B.C.) into his own philosophical system. In what follows, I shall usually refer to Aristotle's theory, although the chemical ideas were originally those of Empedocles.

Aristotle's Theory

In one of the greatest intellectual syntheses of all times, as part of a complete system of logic, metaphysics, physics, biology, psychology, politics, and ethics, Aristotle, in his treatise *Meterologica*, explained the composition of matter, chemical and physical behavior, burning, ripening, decay, putrefaction, color, and geological formation. He postulated that a prime matter exists, and that it is associated with four different properties, or qualities: hot, cold, moist, and dry. Hot and cold are active qualities, and moist and dry are passive. At any given time, only four possible combinations of these qualities can exist in a substance. These combinations are hot and moist, hot and dry, cold and moist, and cold and dry. The other two combinations, hot and cold and moist and dry, are pairs of opposites and cannot coexist in a substance.

Each combination of qualities was associated with an element, the four elements being air, fire, earth, and water. (These were philosophical elements, not the ordinary everyday impure substances.) The relationships between qualities and elements are shown in the diagram below.

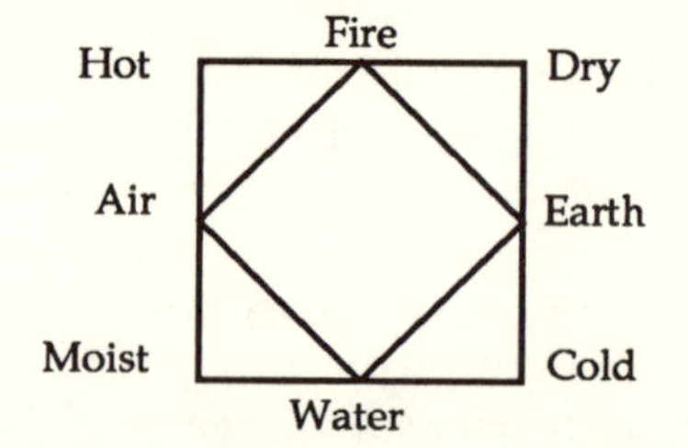

All the different substances on earth were considered to be composed of different combinations of the qualities, and therefore of the elements. The heavens, however, were made up of a quintessence, an element different from and superior to those of the earth. The observed properties of a substance, such as color, viscosity, hardness, and density, were secondary qualities that depended on the particular combination and proportion of the pair of primary qualities present in the material. For example, hard substances had an excess of the quality dry, and dense materials an excess of the quality cold.

All chemicals, according to Aristotle, were made of the same basic qualities and elements. Aristotle used the terms *mixis* and *krasis* to indicate what we call chemical combination. A mechanical mixture was called *synthesis*. When arsenic and sulfur were combined to form the substance realgar, for example, the qualities in the two reactants either combined with each other, or, if they were antagonistic, annihilated each other. The cold in the arsenic combined with the cold in the sulfur, the dry in the arsenic combined with the dry in the sulfur, and so on. To Aristotle, the realgar, which we call arsenic sulfide, was a new combination of qualities and elements, the sum total of which had been in the separate reactants. Because this new combination of qualities was different from that in either the arsenic or the sulfur, those two materials no longer existed. Whenever two or more substances reacted to form products, the original reactants ceased to exist, and each resulting product consisted of a new, homogeneous material completely different from the starting materials. This theory implies that there is no such thing as internal composition or structure and also that substances react with each other in all proportions so that we need not measure out the proportions of starting materials.

Generalizing further, if the only difference between chemical substances is the relative proportion of qualities, any substance can be turned into any other substance just by changing the ratio of qualities. In the Aristotelian system, all reactions were theoretically possible. Furthermore, each element consisted of prime matter only, so it could change or be changed into any of the other three elements. Each chemical substance, therefore, had a composition that could change spontaneously from one moment to the next and so could its observed properties, including mass.

Aristotle considered the qualities to be primary and the elements only of secondary importance. This point of view probably resulted from his own scientific background. He was a biologist, not a chemist or an artisan. He almost certainly never performed chemical operations and had little knowledge of or interest in the different kinds of matter. He did have extensive experience classifying plants and animals on the basis of shape, color, and other observable properties, or qualities. To him, form was function; the organ or specimen behaved in a characteristic fashion because of its form, not because of the matter of which it was composed. On the other hand, those who came after him—the alchemists and chemical technicians, the physicians and chemists, people who were interested in chemical transformations—believed matter (the elements) was of prime importance, not the qualities. They were interested in how matter behaved and what one could do with it. They also recognized that there were simply too many cases of different batches of a particular substance having different shapes, colors, or textures but remaining recognizably the same material. Aristotle's successors soon changed their approach to emphasize the elements. However, their elements, earth, air, fire, and water, are not our modern chemical elements, each of which is an entity different from all other elements. The Greek elements were four interconvertible forms of matter that were believed to be present in all chemical substances.

In light of present knowledge, the ideas of Aristotle and Empedocles were wrong, but in the light of the knowledge of 500 to 300 B.C., they were reasonable, useful, and in agreement with common sense and observation. Moreover, the theory had been built up from observed facts carefully, step by step, by the inductive method[1]. One typical set of the phenomena from which Aristotle's theory developed was the boiling and freezing of water. When water is exposed to cold, it changes to a solid, but when the ice is placed in contact with fire, it turns back to liquid. On further addition of fire, the water boils and changes to air. Another example: Quicklime is a solid, but when ordinary liquid water is added to it, a good deal of heat is evolved and the quicklime changes to another solid. Obviously, the element water has been gained and the element fire has been lost. Other examples are plentiful. Olive oil hardens on standing, the internal water changing to earth. When solid gypsum is heated, fire is added and a different solid, plaster of Paris, is formed. Perhaps the most convincing example was the burning of a green stick. (This demonstration dates from Hellenistic times or later, because it postulates the simultaneous presence of all four elements or qualities, which Aristotle specifically stated to be impossible.) When a green stick burns, it gives off flame (fire) and smoke (air) while sap (water) runs out the end that is not burning. Finally, a solid residue (earth) is left.

---

[1]Aristotle invented the inductive method, maintaining that the basis of all knowledge is experience, that is, sensory perception, and proposing that a general explanation should be induced from the observed phenomena. In other words, he formed his theory starting with the observed facts.

Aristotle was able to explain a whole catalogue of natural phenomena such as melting, burning, fermentation, freezing, evaporation, condensation, the presence of salt in sea water, and the coagulation of milk. Moreover, his theory offered an explanation for the occurrence of underground deposits of metals and minerals. Everyone has observed that when substances are heated, as in cooking, the heat of the fire produces new materials. So in a similar fashion he said the heat of the sun acting on the earth produces two exhalations, a dry hot smoky exhalation and a moist cold vapor. The first exhalation forms minerals in the earth, and the second forms metals.

The four qualities or four elements provided a commonsense explanation of the chemical information of the times with a logical completeness that was not only intellectually satisfying but also aesthetically pleasing to the philosophers of that and succeeding eras.

The Atomic Theory

There were, of course, other theories, most notably the atomic theory of Leucippos (c. 440 B.C.) and Democritos (460–c. 370 B.C.), later modified by Epicouros (341–270 B.C.) and Lucretius (c. 95–55 B.C.). According to their atomic theory, there was only one substance, the prime matter, which existed in the form of changeless, indestructible, indivisible atoms of different sizes and shapes, moving back and forth randomly with nothing between them, in effect moving in a vacuum. (Democritos believed that atomic movements were preordained by the gods. Epicouros and Lucretius believed that the motion was random, the result of something like free will, and had nothing to do with the gods.) The properties of materials depended in some unspecified fashion on the size and shape of the atoms. However, other than the name and the idea that matter is composed of indivisible particles, the Greek atomic theory has almost no relationship to our own[2].

Very few Greek natural philosophers accepted the atomic theory. For one thing, atomism offered no explanations for such natural phenomena as boiling and freezing. More important, however, were their religious reasons. The atomic theory was

[2]Greek atomism may very well have had India as its original source. Indian atomism predated Democritos. Gautama, the founder of Buddhism, is traditionally considered to have been an atomist because on his deathbed, in 483 B.C., he is said to have reminded his disciples that death was merely the disaggregation of atoms. The Greeks and Indians were both in contact with the Persians and probably with each other. In fact, Democritos himself is reported to have journeyed to India (3).

frankly atheistic. It was worked out as part of the Epicurean philosophy, a philosophy that would enable the adherent to lead a good life while ignoring the gods. There was no divine purpose or divine order. The atoms simply moved randomly.

Greek natural philosophers did not separate science from religion. Instead, they looked for a rational, unified explanation of the entire universe. Aristotle offered them a universe that was ordered, logical, and purposeful, obeying a divine plan. Everything had been given its properties, presumably by the gods, for a purpose. Each object was shaped or given its proper allotment of earth, air, fire, and water so as to fulfill that purpose. In line with his concept of the divine origin of all objects, Aristotle postulated that all things, including rocks and metals, had souls. The deposits of rock and metal that were found in the ground lived and grew there, again according to some plan. To understand the purpose of an object or a phenomenon was to understand the object or the phenomenon. Much of Aristotelian natural philosophy was teleological.

The human mind, especially the scientific mind, has always looked for order and purpose in life. Aristotle's comforting system was accepted not only by the Greeks, who believed that nature was ordered and not random, but also, centuries later, by medieval Christian, Jewish, and Moslem theologians and philosophers, who believed that nature reflected God's will. Epicouros and his atheistic ideas were simply not acceptable (4). In Hebrew, not by accident, the word for atheist is *apikoros.*

The Shortcomings of Aristotle's Theory

Neither Aristotle's theory nor the atomic theory of Democritos was completely scientific in our sense of the term. They were based on observation and experimentation but not on experimental testing of premises. There was no attempt to disprove a hypothesis as a test of its validity.

Aristotle relied fundamentally on his common sense. When he looked around him, he saw and touched many different solids, many liquids, and air. Was each solid a material different from all other solids? Was each liquid different from all others? To an educated Greek gentleman who believed that nature was logical, harmonious, and geometrical, there could be only one satisfying answer to these questions. Just as in geometry there were only a few basic propositions from which all others followed, so there had to be only a few basic kinds of matter, one solid, one liquid, and air.

Liquids and solids were obviously different from each other because liquids were moist and solids were dry. But there had to be more to liquids and solids than just moistness and dryness. When heated, liquids evaporate, becoming air. Liquids, therefore, were cold as well as moist, but air was hot and moist. Fire, which was obviously hot, was extinguished by water, so fire was hot and dry. Solids are transformed by heating, so solids are cold as well as dry. The four qualities, hot, cold, moist, and dry, explained the four elements, fire, air, earth, and water. Each substance had different proportions of the qualities and elements. Aristotle could therefore explain all physical and chemical properties on the basis of the four qualities.

Because Aristotle's theory was logical and based on shrewd common sense and observation, it is hard to see how anyone could have done better or even as well with the information then available. Indeed, it was such a success, it seemed so complete and furnished so many plausible explanations for observed phenomena that it shut off debate for almost two thousand years. To anyone trained in Aristotle's system of natural philosophy, which meant any educated person before the seventeenth century A.D., what we today consider chemical research would not only have been meaningless but perhaps a little crazy. Nevertheless, Aristotle's ideas of chemical composition and combination were wrong, and his theory ultimately led into a blind alley.

Perhaps the best way of expressing this is to contrast our definition of chemistry with Aristotle's beliefs. We define chemistry as the study of the internal composition and structure of matter; to Aristotle there was neither internal composition nor structure. As a result, his concept of a chemical reaction was completely different from ours. For example, we say that diatomic molecules of hydrogen and oxygen combine to form water, that the water formed consists of a mass of triangular molecules with empty spaces between them, and that each molecule contains two atoms of hydrogen and one of oxygen. We picture the reaction as follows:

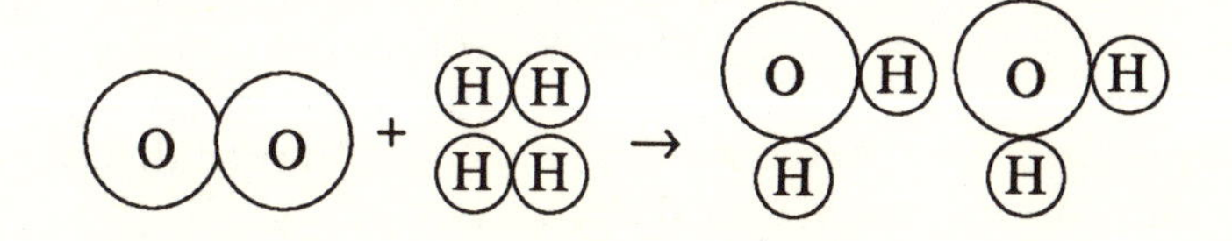

Aristotle, on the other hand, would have considered that the hydrogen, oxygen, and water were each completely homogeneous, because there was no such thing as empty space. Furthermore, both the hydrogen and oxygen disappeared in forming the water. He would have pictured the reaction as follows:

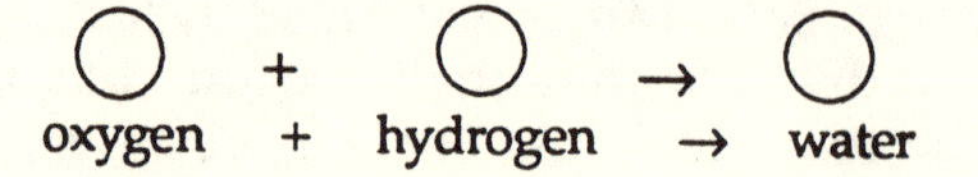

To us, the difference between one material and another depends on the number and kinds of atoms present and on their arrangement. To Aristotle, there were no atoms, and in consequence there was no arrangement. All substances were completely homogeneous.

We know atoms are indestructible under ordinary conditions and no element can be changed into another. To Aristotle, all the elements constantly changed back and forth into each other, and so even if there were atoms, they would not be indestructible.

To an Aristotelian natural philosopher, a program of modern chemical research would have been bewildering. We say that research in chemistry requires purified substances of known composition. This simple statement alone would have confused a Greek philosopher. It contains concepts that he would not have understood and would have rejected once he understood them.

What do we mean by a pure substance? What do we mean by composition? We define a pure substance in terms of operations by saying that it has a constant composition, that the chemical analysis does not change when it is purified further. For example, in pure water there is 1.00 gram of hydrogen for each 8.00 grams of oxygen. By constant composition we imply that on chemical analysis each pure substance will have definite weight percentages of the constituents. For example, 232 grams of cinnabar (mercuric sulfide) will, if pure, contain 200 grams of mercury and 32 grams of sulfur. To Aristotle and his followers, however, cinnabar was just cinnabar. It did not actually contain either sulfur or mercury. Analysis for sulfur and mercury would be looking for things that were not there.

In a modern chemical investigation, once the starting materials have been purified, the chemist studies the reaction or reactions to see the effects of the shape and composition of the reacting molecules. To Aristotle, there were no molecules, no composition, and, of course, no shapes.

Finally, one of the most basic principles of modern chemistry is the conservation of mass. The chemist must end a reaction with the same mass of material (including any gases) that he started with—or he has made some egregious blunder. To Aristotle, although the conservation of matter was logical and fundamental, the conservation of mass would not only have been wrong in theory but also demonstrably false in practice. Mass could not possibly be conserved because earth and water, which were heavy, were constantly being converted into fire and air, which were light, and vice versa. After all, many substances were observed to change weight spontaneously. Quicklime got heavier on standing. Wine got lighter. When impure gold was heated in a refiner's fire, it lost weight. Aristotle could easily have cited many cases where the visible products of a reaction were either heavier or lighter than the reactants.

For modern chemistry to develop, the entire Aristotelian frame of reference had to be abandoned. The chemist had to think of chemicals and reactions in a completely different way. It took two thousand years for this way of thinking to emerge.

References

1. Farrington, B. *Greek Science*; Penguin: Harmondsworth, Middlesex, England, 1944; p 24.
2. Warbasse, J. P. *The Doctor and the Public*; Hoeber: New York, 1935; p 69.
3. Partington, J. R. *History of Chemistry*; Macmillan: New York, 1970; Vol. I, p 29.
4. Multhauf, R. *The Origins of Chemistry*; Franklin Watts: New York, 1966; pp 50–51.

Timeline—Hellenic Chemical Science

	600–550 B.C.	550–500 B.C.	500–450 B.C.	450–400 B.C.	400–350 B.C.	350–100 B.C.	100 B.C.–1 A.D.
Technological and Philosophical Advances	Thales predicts solar eclipse in 583 B.C.	Leucippos develops first atomic theory, c. 530	Empedocles develops Four Qualities, Four Elements theories	Democritos (460–370) revises atomic theory; Greek alphabet standardized by Athenians; Plato (428–348)		Epicouros (341–270) proposes atomic theory as part of an atheistic philosophy	Lucretius writes *De Rerum Naturae*
Historical Landmarks	Greek Ionian cities flourish c. 600	Cyrus the Great, first Persian emperor, c. 530; Persians conquer Ionia, 525	Greek–Persian War, 490–449	Pelopponesian War, 431–404			

III

The Hellenistic and Roman Eras

The Rise of Alchemy

SINCE ANCIENT TIMES, people have worked with chemical materials of one sort or another. And from about 600 B.C., and probably even today in some parts of the world, people have been performing what is now known as alchemy. However, most of those working with chemicals were neither alchemists nor chemists but artisans and craftsmen.

The word alchemy first appears in documents around the fourth century A.D., written as *khemia.* Perhaps it was derived from the Egyptian word *kmt* or *kemi* (*1*), "the black land" (Egypt). Or it might have come from the Greek word *chyma* referring to melting and casting metals (*1*). A Chinese origin is also possible (*2*). An old Chinese word for gold-making sounded like *kemi* and could have reached the West through the Arab sailors and merchants who traded with towns on the South China coast. In any case, there seems to be no doubt that the Arabs added the prefix *al*, "the", to form *al-chemy* or *alchemy.*

Although it is clearly related to alchemy, early chemistry is not descended from it. The ideas of alchemy were alien to those of chemistry, and the one did not evolve into the other. Their relationship, instead, is like that between humans and chimpanzees, two entirely different species that are related and somewhat similar in appearance. Even though both alchemy and early chemistry used many of the same substances and processes and much of the same equipment, their aims were different. Alchemists were concerned primarily with finding potions to produce gold or confer immortality (the exoteric alchemists) or to save souls (the esoteric alchemists). Chemists, on the other hand, were and are interested in using chemicals to make other chemicals that have an enormous variety of everyday functions. Chemists, like alchemists, want to be rich, live a long time, and go to heaven, but they go about their aims differently.

Hellenistic and Roman artisans were at least as capable as their predecessors, and they invented new and important processes and materials. They produced metals and alloys, jewelry, glass, cements, dyes, pigments, artificial pearls, gems, and medicines. Pliny (23–79 A.D.) reported the first use of indicator paper, in the form of papyrus strips soaked in extract of oak galls, to detect iron as an adulterant of blue vitriol (copper sulfate). The early Mesopotamians had relied on naturally occurring asphalt and bitumen for mortar in laying bricks, although they did sometimes use artificial cements to bind the bricks together. By the time of the Roman Empire, there were some twenty-eight different kinds of cement, each being applied in specific circumstances. A Pompeian wall painting made before 79 A.D., the date of the great eruption of Vesuvius, provides an example of Hellenistic technology and social attitudes. In the illustration on the next page, a wool bleacher is shown carrying a pot of sulfur and a hemispherical wicker frame. The woolen cloth was spread out on the wicker frame, and the sulfur placed underneath and ignited; the sulfur dioxide fumes bleached the cloth. Wool bleachers, constantly exposed to the corrosive sulfur dioxide fumes, must have had short, pain-racked lives. Linen bleachers were perhaps even worse off. Linen was bleached in alkaline solutions in which the bleachers sometimes had to tread or knead the cloth, in which case the alkali would eat holes in their skins (3). After being saturated with alkali, the linens were spread out in the sunlight to dry. The light, acting on the alkaline solution in the cloth, generated hydrogen peroxide, which did the bleaching. Cotton, too, was bleached, but with an acid peroxide solution. It was dipped in sour milk (a dilute solution of lactic acid) before being spread out in the sunlight.

Early Alchemical Literature

This chapter is concerned primarily with alchemy, rather than directly with technology. It is almost impossible to trace the ori-

Roman wool bleacher with equipment. (Reproduced with permission from Oxford University Press.)

gins and growth of alchemy. The most difficult problem is the lack of reliable documentation. The early alchemical doctrines were transmitted orally from master to disciple and very little written material exists that can be dated with any degree of confidence to this early period. What documentation we do have seems to have been written down centuries after the rise of alchemy. Most of the extant manuscripts were copied from other, lost, papyri. Inevitably errors of omission and commission have crept in, so that by now even descriptions that were perhaps originally straightforward have become garbled. Moreover, the writings often used symbols and code words so as to be intelligible only to other alchemists. Even descriptions of nonalchemical processes were concealed with false names and false instructions because they contained trade secrets.

Some early alchemical works were obviously intended as recipes. The Leiden Papyri of the third century A.D. contained one for fake gold:

> To give objects of copper the appearance of gold so that neither the feel nor rubbing on the touchstone can detect it, to serve especially for a ring of fine appearance. Here is the process. Gold and lead are reduced to a fine powder like flour, two parts lead to one of gold. When mixed, they are mixed with gum and the ring is covered with this mixture and heated. The operation is repeated several times until the article has taken the color. It is difficult to detect because rubbing gives the mark of a genuine article and the heat consumes the lead but not the gold [*4*].

This is an early appearance of deliberate deception by the alchemist. The process is actually gold plating by applying a lead–gold mixture and evaporating off the lead. French goldsmiths of the sixteenth to nineteenth centuries used a similar process with mercury instead of lead to make bronze doré ornaments.

The Stockholm Papyrus contains a recipe for the field of costume jewelry:

> To make artificial pearls . . . mordant or roughen crystal in the urine of a young boy and powdered alum, then dip it in quicksilver and woman's milk [*5*].

Here "quicksilver" does not mean mercury but probably a suspension of fish scales in water. Almost certainly "the urine of a young boy" was a code name for something else as was the "woman's milk."

Another alchemical recipe was for fake emeralds:

> Take white lead, one part, and of any glass you choose, two parts, fuse together in a crucible and then pour the mixture. To this crystal, add the urine of an ass and after forty days you will find emeralds [*6*].

Stillman considered that this was a recipe for green glass, the green color coming from a copper salt, code named "urine of an ass" (*6*).

One final difficulty in understanding the alchemical literature is that the viewpoint of the alchemist was so different from that of today's chemist that it is almost impossible for us to be certain of what is being stated or described. To a considerable extent, therefore, the interconnections between the real and the mythical alchemists and the various alchemical activities are conjectural. Much of what follows here, unless it is clearly stated otherwise, is simply my opinion based on my interpretation of standard sources.

Allegorical representation of lead as a slow, crippled old man (because of lead's density). The representation is a woodcut dating from the Renaissance. (Reproduced with permission from the Bettmann Archive.)

Esoteric Alchemy

It is generally agreed that there were two major divisions of alchemy. Esoteric, or religious, alchemy was an attempt to understand God, or the gods, and to find salvation. Exoteric alchemy, on the other hand, was definitely worldly, and sometimes even crass. In the Near East, exoteric alchemy was the search for the philosopher's stone, which would transmute base metal into gold. From very early times in China and somewhat later in the Near East and Europe, exoteric alchemists were also interested in the elixir of life, a potion that would enable the user to live forever. The search for the elixir contributed to medical alchemy, which influenced the development of both medicine and chemistry. The search for the philosopher's stone contributed little to chemistry but did affect the popular imagination—the seeker of gold being the person we usually mean when we use the term *alchemist*. The gold-seeker is the alchemist shown, centuries later, in various paintings and etchings of laboratories and satirized in Chaucer's *Chanoun's Yemanne's Tale* and Ben Jonson's play, *The Alchemist*.

However, the deliberately obscure esoteric alchemists contributed most to what eventually became chemistry. They considered chemical transformations as analogies and used their observations to fashion religious allegories. It was essential, for religious reasons, for them to keep careful records of what they saw and did. They collected and recorded their results, although in cryptic and disguised form. This information ultimately found its way to succeeding generations of alchemists, artisans, physicians, and natural philosophers.

We must be careful, however, not to make too rigid a distinction between artisans and esoteric and exoteric alchemists. They were not three distinct groups. Often they were the same people practicing, at various times, three somewhat different activities. After all, alchemists usually earned their bread by working as artisans, and both artisans and alchemists almost certainly attempted to make gold at every opportunity. Moreover, both artisans and exoteric alchemists believed in much of esoteric alchemy.

Esoteric alchemy came into existence before exoteric alchemy, probably as an offshoot of the mingling of astrology with religion. Astrologers believed, and still believe, that events on earth are governed by the heavens. They studied the heavens to understand the earth, and some of them took the logical position that one should also study earthly events to understand the heavens. To them the heavens (the macrocosm) and the earth (the microcosm) were mirrors of each other; one could understand either by studying the other.

Among the earthly events particularly studied were those involved in metallurgy. Early societies had always believed that rocks and other inanimate objects were alive, possessed of intelligence and magical powers. Even Aristotle believed that metals and minerals had souls and grew in the ground. (We still encounter the term "the living rock.") The reactions taking place in the furnace were obvious magical acts of creation with religious analogies. (In the eighteenth century, Handel wrote, in *The Messiah*, "For He is like a refiner's fire.") Astrology and esoteric alchemy became inextricably mixed with each other and with mystical religion.

After about 600 B.C., there developed all over the ancient world a deeply felt desire for personal salvation and eternal bliss, perhaps as compensation for the miserable conditions under which the mass of humanity existed. Personal religions arose, including Buddhism, Mithraism, Christianity, Manichaeism, and, later on, Islam. Astrologers and alchemists were among those interested in their immortal souls, and salvation rapidly became the purpose of alchemical study[1]. Man became identified with the microcosm as summed up by Protagoras (c. 485–410 B.C.), who stated unequivocally, "Man is the measure of all things." Eventually, the mixture of astrology, metallurgy, and alchemy became a cult, or several cults, which we call by the general term esoteric alchemy.

These cults included the Moslem Brethren of Purity (or the Faithful Brethren) in Africa and the Near East and Taoism in China, both deeply involved in alchemy. In Europe the major alchemical cult was Gnosticism. There is still much controversy about its origins and specific doctrines, but probably it began in Mesopotamia as a mixture of Babylonian, Egyptian, Persian, Greek, and, later on, Christian beliefs. Its fundamental idea was that knowledge was salvation. Gnosticism was most influential from 100 to 600 A.D., after which it was suppressed by the Christian authorities. Even so, although officially condemned, gnostic ideas were preserved in the alchemical literature of the Christian era. They seem to have influenced thinkers as diverse as the medieval mystic and healer, Saint Hildegarde of Bingen (1098–1179), and the sixteenth century medical alchemist, Paracelsus. Paradoxically, in medieval Europe, even though esoteric alchemical cults were violently suppressed by the Church, alchemical theory in support of Christian theology remained respectable. Saint Thomas Aquinas, for example, studied alchemy with Albertus Magnus and used alchemical ideas in his works.

Understanding and interpreting alchemical writings are complicated by their possible religious and mystical significance. A cryptic passage that is ostensibly a recipe for a chemical process might actually be a religious tract. Here, for example, is a translation of a fragment from Bolos of Mendes, who probably

[1]"The hermetically sealed retort, in which were placed the mystical metals, was regarded as a realm apart—a special region of heightened forces comparable to the mythological realm; and therein the metals underwent strange metamorphoses and transmutations, symbolical of the transfigurations of the soul under the tutelage of the supernatural" (7).

lived between 300 and 200 B.C. It seems to refer to the transmutation of base metal to gold, but there are symbols, false names, and secret codes. It could mean almost anything and, to say the least, is certainly obscure.

> Here is the mystery: the serpent Ouroboros, this composition which in its ensemble is devoured and melted, dissolved, and transformed by the fermentation or putrefaction. It becomes a deep green and the color of gold is derived from it. It is from it that is derived the red called the color of cinnabar. This is the cinnabar of the philosophers. Its stomach and back are the color of saffron, its head is a deep green. Its four feet constitute the tetrasomie. Its three ears are the three sublimed vapors. The One furnishes the Other its blood and the One gives birth to the Other: nature rejoices in nature; nature triumphs over nature; nature masters nature; and that not for a nature opposed to such another nature but for one and the same nature proceeding of itself by the process with trouble and great effort. But thou, my dear friend, apply thy intelligence to these matters and thou wilt not fall into error; but work seriously, and without negligence, until thou hast seen the end. A serpent is stretched guarding this temple and he who has subdued it commences by sacrificing it, then roasts it, and after removing its flesh up to the bones makes of it a step to the entrance of the temple. Mount upon it and thou shalt find the object sought, for the priest at first a man of copper has changed color and nature and has become a man of silver; a few days later, if thou wish, thou wilt find him changed to a man of gold [*8*].

Influence of Greek Science

Both esoteric and exoteric alchemy were directly affected by contact with Greek science, which spread throughout the conquests of Alexander the Great. Alexander's empire stretched from the Adriatic to India and included Egypt, Syria, and Mesopotamia, the great centers of technology, and Greece, the center of philosophy. At his death in 323 B.C., his domain broke up into three major warring states: Egypt, Macedon (which dominated Greece), and the Seleucid Empire (which included Mesopotamia and western Persia). Syria was disputed territory, fought over by both the Seleucids and the Ptolemies, who ruled Egypt.

Both Seleucids and Ptolemies relied on Greek soldiers and a Greek bureaucracy to rule over a subject native population. It was relatively easy to recruit Greeks because their state was now a stagnant backwater, and ambitious Greeks left their homeland for careers in Mesopotamia, Syria, and Egypt. To encourage this influx of talent for the army and bureaucracy, the Ptolemies and Seleucids adopted a policy of hellenization. Among other expedients, they set up centers of higher learning to attract philoso-

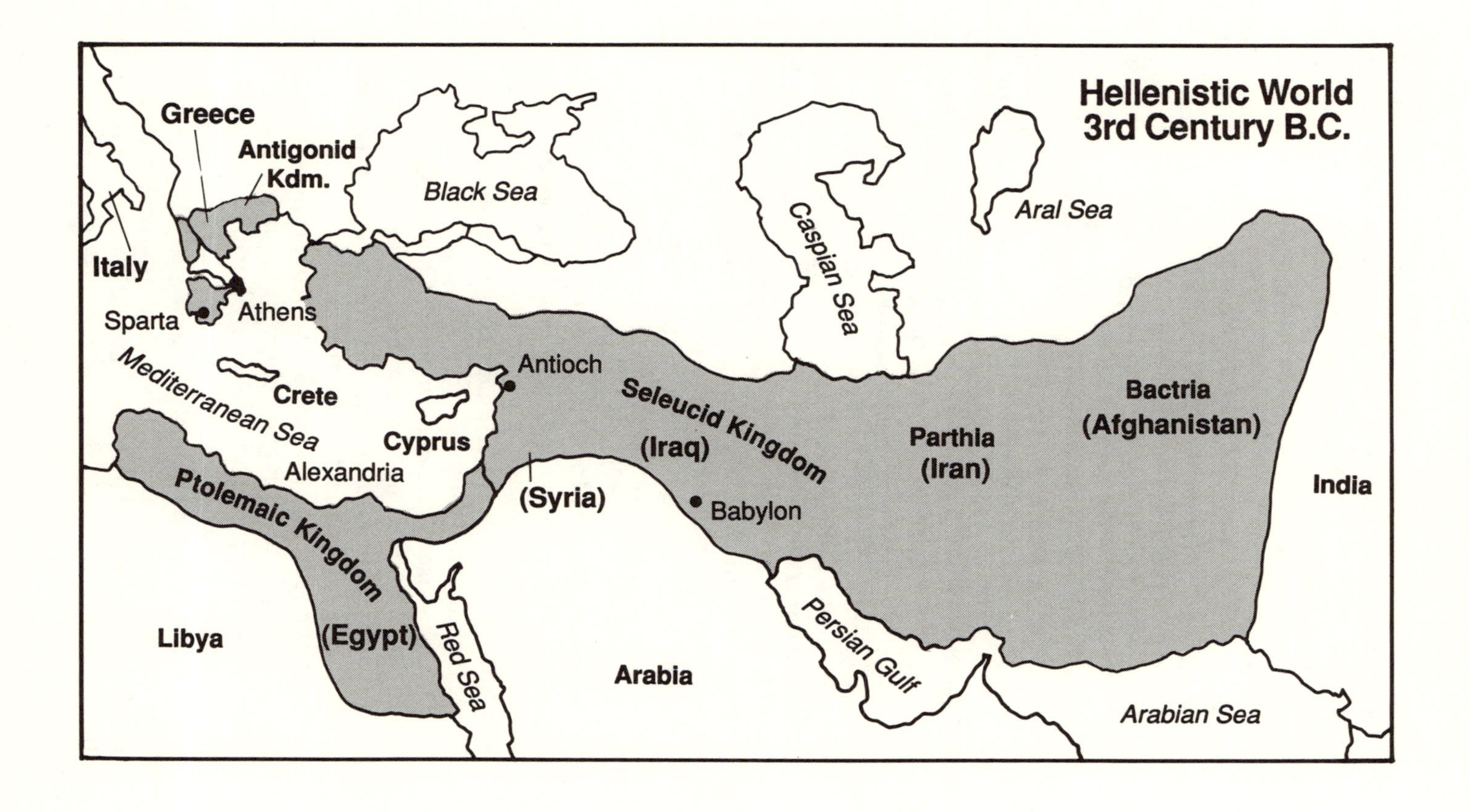
Hellenistic World
3rd Century B.C.
Greece
Antigonid
Kdm.
Black Sea
Caspian Sea
Aral Sea
Italy
Sparta
Athens
Antioch
Mediterranean Sea
Crete
Cyprus
Alexandria
Seleucid Kingdom
(Iraq)
(Syria)
Babylon
Parthia
(Iran)
Bactria
(Afghanistan)
India
Ptolemaic Kingdom
(Egypt)
Libya
Red Sea
Arabia
Persian Gulf
Arabian Sea

phers and scholars. This encouragement continued the tradition of Alexander, who had been the pupil and friend of Aristotle, for whom he collected biological specimens even during his campaign against Persia.

The greatest of the scholarly institutions was the Museum at Alexandria. Bitter and prolonged fighting took place in Asia Minor, Mesopotamia, and Persia throughout the Hellenistic period, from Alexander's death in 323 B.C. until the rise of Rome. Cities were sacked, libraries and museums destroyed, and scholars killed or dispersed. Egypt alone escaped the effects of war and pillage, and for hundreds of years the Museum at Alexandria was the center of learning of the Mediterranean and the Near East. Attracting scholars from all over the Hellenistic world, the museum, together with the Serapion, the library of the Temple of Serapis, became an enormous university. The museum library alone had more than a half million parchment scrolls, and the Serapion had another half million.

For all the large number of scrolls in Alexandria, however, Greek science reached the Egyptians and Syrians by word of mouth. Books were scarce and expensive. They were not meant to spread information but to preserve it. Greek natural philosophy was carried by the wandering Greek philosophers. They would journey to Antioch, Alexandria, or any of the smaller Ptolemaic or Seleucid cities and lecture to the crowd in the agora, a combination marketplace and civic center. After their speeches, they would collect whatever donations they could—perhaps even an offer of free room and board. Those who were received enthusiastically would stay for an indefinite period, teaching disciples for a fee. If not welcomed, they would move on. When they lectured in the crowded agora, fellow Greeks listened to them and so did many native inhabitants who understood Greek. These included not only the Egyptian and Syrian merchants but also many artisans and craftsmen who had shops nearby. And when the organized natural philosophy of the Greeks reached the jewelers, smiths, and alchemists, it fell on receptive ears.

Aristotle's doctrine that metals and minerals had souls made sense to the esoteric alchemists, and from it they drew a logical conclusion: Distillation and sublimation purified material substances, so these processes obviously purified the souls of these substances—and by extension could purify human souls as well. Early on, esoteric alchemists became preoccupied with distillation and sublimation. (The word sublime originally referred to the solid material purified by evaporation and recondensation.) In distillation, the residue left in the flask was termed the dead body or *caput mortuum* (literally dead head), and the condensed vapors were considered to be the soul or spirit of the material. The term "spirit" carries over even today in the names of such pharmaceuticals as spirits of ammonia and sweet spirits of niter.

Exoteric Alchemy

Greek science stimulated exoteric alchemy too. Probably ever since the first metals were produced from rocks, jewelers and smiths had tried to make gold from minerals, metals, and whatever else they could think of. Eventually, discouraged by repeated failures, most artisans who tried it gave up. After all, there was no reason to believe that it could be done, although in the absence of any coherent theory there also was no reason to think it couldn't be done. Greek science, however, changed that view.

When the jewelers and smiths heard of the four qualities and four elements, it immediately became obvious to them that gold really could be made from lead and other base metals. All they had to do was to change the proportions of the qualities from those of base metal to those of gold. In theory, it was possible. The only problem was practical, the choice of the best procedure to follow. The search for this procedure is what we call exoteric alchemy. From about 300 B.C. until at least the seventeenth century A.D.[2], exoteric alchemists searched both extensively and intensively for the way to transmute metal into gold.

The procedures adopted by the exoteric alchemists inseparably intertwined alchemy and magic. The connection was inevitable because transmutation was believed to be a process of growth, reproduction, and death. Fertility rites were involved in such processes, and practitioners recited ritual formulas, used materials of occult and sexual significance, such as snake venom and rooster combs, and perhaps even offered sacrifices. Also sulfur and mercury, which had religious and magical significance, were very much involved in the transmutation procedures. The Greek word for sulfur was *theion*, which meant "divine" as well as "sulfur" (*9*). The divine sulfur had dual significance in the alchemical process. Even a small amount of sulfur added to metal produced dramatic color changes, and moreover, sulfur had always been associated with magic. Sulfur is the biblical brimstone, and sulfur and sulfurous odors are evolved in volcanic eruptions and sometimes in earthquakes and lightning strikes.

[2]Even in the seventeenth century, the hard-headed Cardinal Richelieu inquired into having an alchemist produce enough gold to help pay the costs of the Thirty Years' War.

Witches and demons had a sulfurous smell, and the damned were believed to burn in sulfur fires in hell. Mercury too was magical because it was quicksilver, a liquid that did not wet anything and was impossible to pick up.

The theoretical basis for alchemical transmutation procedures was found in Aristotle's works. Aristotle had stated that metals and minerals grow in the ground, like plants, although not as rapidly[3]. Growth is the attempt to reach perfection. Perfection of metals was represented by incorruptible gold. Impure metals grew slowly in the ground and turned into gold over the course of centuries. Looked at in that light, gold was fully ripened metal. Therefore, all the alchemist had to do to obtain large quantities of gold was to start with a batch of impure metal and speed up the ripening process.

Young plants and animals grow faster than old ones, so by analogy, young metals should grow and ripen faster than old ones. Therefore, for rapid transmutations, the alchemist should start with young metal or else somehow make the metal young. Again reasoning by analogy, just as young plants are formed from seeds, so young metals should be formed from metal seeds. Among plants and animals, the individual dies but its seed reproduces the species. So too with metals; individuals die and the seeds germinate into new metal. Putting all these ideas together, the alchemists worked out a theoretical process for speeded-up transmutation that involved the death of a mass of metal and its rebirth and rapid growth from seed.

Each alchemist had his or her own set of formulas and spells, rites and incantations, methods of mixing and heating materials, and preferred astrological conjunctions. But the basic themes were the same. First the starting material in the transformation was killed to break it down into prime matter, which would nurture the metal seed. Then a seed was introduced and encouraged to grow and produce the gold. Because colors were characteristic of life and states of being, the different stages in the process were monitored by the color of the metal.

To kill the metal, alchemists either heated it at moderate temperatures to oxidize it, or treated it with sulfur or sulfide solutions. The proof of death was the loss of metallic properties and a blackening (*melanosis*) of the surface. Once the metal was dead, the seed was introduced. Originally the seed was probably

---

[3]This belief persisted in western Europe, even in scientific circles, as late as 1800. There is extant a letter from a high official in the service of Louis XIV instructing his subordinates to seal up the entrance to a depleted gold mine so that it could renew itself. At about the same time, in Spain fifteen years was believed to be the amount of time a mine needed to regrow its deposits (*10*).

a small piece of gold that was intended to grow and increase in size. Later on, the concept was modified, and the seed became what we would now call a catalyst. It induced changes in the metal, but it did not itself change. The new catalytic seeds were not pieces of gold. Instead, they were coloring agents, either liquids or powders, called tinctures and iksirs. The latter term became more widely used. Later on, in the Arab era after 650 A.D., Arabs added their usual prefix, *al-*, and it became al-iksir and, in English, elixir. About the same time, possibly as a result of Chinese influence, alchemists also began to use the term philosopher's stone. Originally this was supposed to be a panacea for human illness, but the idea was transferred to the transmutation of base metals. Metals other than gold were sick, and the philosopher's stone cured them, changing them to gold.

Once the seed had been introduced, the metal was heated at a semi-controlled temperature for a prolonged period, and thus the gold was grown. At the proper time, the mixture was treated with orpiment (arsenic sulfide). The orpiment produced a whitening (*leukosis*), signifying that the metal had now been converted to silver, halfway to gold. Finally the metal was treated with a solution of polysulfides, called *theion hudor*, which meant both "divine water" and "sulfur water." This procedure, called *xanthosis* (yellowing), gave the metal the appearance of gold.

Transmutation was lengthy, laborious, and expensive. The heating period might take days or weeks, during which the heating had to be controlled and the mixture watched. Any minor slip, a mistake in the rites when adding a reagent, preparation of a reagent under the wrong planetary influences, anything, could (and, of course, always did) ruin the entire transmutation. But the alchemist would never know until the lengthy process was over whether or not he or she[4] had succeeded. And when it failed, there was no way to know what had gone wrong.

In any event, at the end of his labors, the alchemist had a mass of yellow or yellowish metal. He knew, of course, that not all of the yellow product was gold, but he believed surely some of it was. He had to find out how much was really gold and how much was just base metal dyed yellow, and the only way to do so was by *cupellation*, that is, analysis by fire. (True gold was incorruptible; no known reagent could attack it, and even the intense heat of the furnace had no effect.) The Hellenistic alche-

[4]There were women alchemists too, including Cleopatra; the legendary Miriam, the sister of Moses; and Mary the Jewess, who is supposed to have invented the laboratory water bath, which the French still call the bain-marie and cooks call the double boiler.

mists, who were much better educated than the ancient craftsmen, tested their product scientifically by cupellation to find out how much was gold. Ironically, the results of their scientific tests misled them into believing that they had indeed made gold.

Before 1500 B.C. in ancient Mesopotamia, cupellation had been developed to refine gold by removing any base metal[5]. The impure gold or silver is placed in a porous porcelain cup, called a cupel, and is heated to a high temperature while a stream of air blows over its surface. The oxygen in the air forms oxides with any reactive metal present, but not with the unreactive silver and gold. When the heating is finished the cupel is cooled and examined. All the base metal is now gone, any oxides having been absorbed by the porous porcelain of the cupel. Any gold or silver present in the hot mass is only melted, not oxidized, and when cooled, remains in the form of a gleaming button of precious metal at the bottom of the cupel. The cupellation method, although accurate, is slow, tedious, and expensive, because it requires the labor and expense of making the cupels and keeping the furnace hot for a prolonged period.

When the Hellenistic alchemist finished transmuting a batch of lead and tested the product by cupellation, he or she found gold and silver present, although in small quantities. That convinced him that he really had transmuted lead into gold or silver. But actually he had done no such thing. Small amounts of gold and silver are almost always present in the mineral deposits from which lead is obtained. In fact, it would have been unusual for an alchemist to "transmute" lead and not find any gold or silver.

This proof of transmutation was no proof at all. It would have been proof only if the alchemist had run a control by cupelling a blank, a sample of the same lead that had not been transmuted. But who in the third or second century B.C. would have thought of doing such a thing? Only someone who doubted the basic theory would go through all the labor and expense of cupellation on nontransmuted lead. But who doubted Aristotle and Empedocles? Certainly not the alchemist, who of all people, wanted Aristotle to be right and passionately wanted to believe it was possible to transmute lead into gold.

Honest alchemists who cupelled their products and did find some gold were now in a trap, not of their own making. They were sure they were on the right track. Complete success was

[5]In the Old Testament, Malachi (3:2,3, 4:1) refers to purification of silver and gold by fire. Also in the Tel-el-Amarna letters, about 1375 to 1350 B.C., the king of Babylon wrote to the Pharaoh of Egypt complaining that cupellation proved the gold sent to him to be impure.

just around the corner. With a little more heat or a larger seed or some minor variation in the magical formulas, they would produce gold on a large scale. So they would repeat their transmutations again and again, spending their savings and anything they could beg or borrow—all to no avail. When their resources were gone, rather than give up the quest, they might even resort to chicanery, counterfeiting gold and selling the counterfeit to the credulous to raise money for more attempts at transmutation.

Eventually it began to dawn even on the alchemists that no matter what they tried, they were getting only very small amounts of gold. People, including government authorities, became convinced that alchemists were charlatans. Already in 292 A.D., Roman treasury officials evidently suspected that alchemists were counterfeiting coins. Emperor Diocletian outlawed alchemy and ordered the alchemical texts burned. Still, between 300 B.C. and 300 A.D. almost everyone believed firmly in transmutation.

Alchemical Contributions

The alchemists made significant contributions to the practice and theory of Greek natural philosophy. Alchemical procedures involved prolonged heating and needed some sort of temperature control. Consequently, furnaces and ovens were improved, and new types of heating baths were invented. Equipment for distillation and sublimation was still inefficient and not much better than that of the ancient Mesopotamian perfumeress, two thousand years earlier. Nevertheless, new types of stills, beakers, receivers, filters, and other pieces of equipment were invented. We still use the term *hermetic seal* for a seal supposedly invented by the legendary Hermes Trismegistos.

The exoteric alchemists especially did more than merely accept Greek doctrine. They modified it in light of their own observations and experience. They were laboratory workers, not speculative philosophers. In dealing with matter, thinking in terms of matter, little by little they changed Aristotle's theories, shifting the emphasis to matter, from the qualities to the elements.

Moreover, they introduced a new type of matter intermediate between the unseen elements and the substances they could see. Among all the sublimations and distillations carried on by exoteric alchemists was a good deal of what we call destructive distillation of organic materials. They heated wood, coal, horn, or other naturally occurring substances in containers so that the fire had no direct contact with the materials being heated. When they did so, the hot wood or other organic matter did not burn but decomposed in a dramatic fashion. Dense yellow fumes were produced, and white smoke formed when hot vapors came into contact with the colder air. When the vapors cooled, they formed liquid. Moreover, some of the vapors could burn. Finally, at the end of the distillation, a mixture of black tar and charred wood was left in the heating vessel.

The alchemists, who watched destructive distillations, could not help wondering about the origin of the vapors and tar arising from wood. The wood being distilled was not in contact with any material, other than the container, and so, little by little, they began to believe that the tar and the vapors had somehow been present in the wood all along and were simply liberated by the heat of the fire. Of course, the crude tar and the vapor were so different from the wood that they could not really have been present in the final form the alchemists saw. They must have been present in the wood as *philosophical essences* or *spirits*, which in impure form looked like tar and vapor. Therefore, the building blocks that made up wood and all other materials that could be destructively distilled were not earth, air, fire, and water but the essences or principles of vapor and tar. These essences, in turn, were made of the basic elements earth, air, fire, and water.

In effect, the alchemists were now explaining the properties and reactions of substances in terms of matter, not qualities—and matter in the form of complicated combinations of the elements, namely, the principles.

The existence of these principles could readily be justified by Aristotle's doctrines. There were both combustible vapors and condensible vapors, so there was both a combustible principle and a condensible principle. These were readily identified with Aristotle's exhalations. The combustible principle was the hot, dry, smoky, masculine exhalation, namely sulfur. The condensible principle was the cold, moist, feminine exhalation, namely mercury, the spirit of fluidity.

A further shift in emphasis arose from the exoteric alchemists' preoccupation with metals. Although at first they accepted Aristotle's postulate that the exhalations acted on earth, air, fire, and water to form minerals and metals, gradually the alchemists decided that the exhalations were themselves constituents of the metals. One metal therefore differed from another only in the different proportions of sulfur to mercury.

By the end of the Hellenistic era, one more important change had occurred. The age-old concept that metals had souls merged with the new idea that metals were made of sulfur and mercury. Sulfur was now identified as the soul of the metal, and mercury was its intelligence. There is no direct evidence for the Hellenistic origin of this idea, but the conquering Arabs referred to it in a manner that indicates it did not originate with them but was already accepted doctrine when they encountered it.

The End of the Hellenistic Era

After about 300 A.D., progress in alchemy and natural philosophy ground to a halt. As the Roman Empire struggled with civil wars and barbarian attacks, society became increasingly divided into the very poor and the very rich. The peasants and small farmers who had made up the Roman armies had been beaten down into what became serfdom. As a result, at the end there were few native Romans willing to be soldiers to defend the empire. The armies consisted mostly of German, Arab, and Hun mercenaries, the very people who were trying to break into the empire. Commerce shrank and with it shrank the merchant and artisan classes. Wealth was concentrated in land, and the wealthy landowners retired from the cities and towns to live on, and defend, their estates. There was a marked decline in the urban leisure class that could afford to be interested in intellectual activities such as natural philosophy.

Then, with the triumph of Christianity in the Roman Empire, classical science came under attack. The early Church was hostile to pagan philosophy and especially to alchemy. From the time of the first Christian Emperor, Constantine (ruled 306–337), pagan temples and libraries were under physical attack. In Egypt in 389 the Temple of Serapis was destroyed, and its library with it. In 412 in Alexandria, the last pagan philosopher of note, the Lady Hypatia, was lynched by a mob of monks. In 415 the Museum at Alexandria was closed, and much of its library scattered. Finally, in 529, the Academy of Plato in Athens was closed after eight hundred years of operation, and the remaining philosophers fled to Persia. Hellenistic science was by now pretty thoroughly destroyed in its homeland. However, some manuscripts were saved and stored in monasteries, especially those of the dissident Monophysites and Nestorians.

In 489 the Nestorian academy at Edessa, Syria, was closed and the Nestorians banished. They moved to the towns of Nisibis in Mesopotamia, and Jundishapur in Persia, where they received the protection and support of the Shah, at least partly because he felt that "the enemy of my enemy is my friend." They translated their collection of Hellenistic works into Aramaic, the language spoken in Syria and Palestine. As fast as new works were sent out or smuggled out of the Roman Empire, these too were translated.

The end of the Hellenistic–Roman era came with the Arab conquest. Between 640 and 720, Egypt, Persia, Syria, North Africa, and Spain all fell to the Arab invaders. The astonishing speed of the conquest was largely a result of the military and economic exhaustion of the Romans and Persians and internal unrest and subversion within their empires. The Roman Empire had split into eastern and western halves. The Eastern Romans (the Byzantines) had fought a long series of devastating wars with the Persians and were completely drained of money and warriors. The governments were hated, especially that of the Byzantine Empire, and when the Arabs attacked, the dissident residents refused to fight. In fact, they often welcomed the Arabs as deliverers. Alexandria, a huge city with an open and impregnable port and a large garrison backed by a fleet that could bring in supplies and reinforcements at will, surrendered almost without a fight to a handful of men who rode in from the desert. Quite simply, the Alexandrians preferred the Arabs to their oppressive Byzantine rulers.

The Arab conquest marked the end of classical times. Except for a small remnant of the Byzantine Empire in Anatolia and the Balkans, the religion, official language, and laws of the Eastern and Southern Mediterranean were no longer based on Christianity and Greek and Roman traditions. Western Europe was cut off from its intellectual base for the next four or five hundred years, with only minor cultural contact, although commerce between the Christian and Moslem worlds never completely stopped. Until the eleventh or twelfth centuries, the great natural philosophers were predominantly Islamic.

References

1. Forbes, R. J. *Studies in Ancient Technology*; Brill: Leiden, Netherlands, 1955; Vol.1, p 122.
2. Mahdihassan, S. *Ambix* **1976,** *23,* 129.
3. Personal communication from R. P. Multhauf.
4. Stillman, J. M. *The Story of Alchemy and Early Modern Chemistry*; Dover: New York, 1960; p 84.
5. Ibid., p 89.
6. Ibid., p 160. Reference to P. E. M. Berthelot, *La Chimie au Moyen Age,* Vol. II, p 29.

7. Campbell, J. *The Hero with a Thousand Faces*; Princeton University Press: Princeton, NJ, 1968; p 73.
8. Stillman, J. M., op. cit., p 171.
9. Aristotle. *Problematica XXIV 19*, 937b, cited by Partington, J. R., *A History of Chemistry*; Macmillan: New York, 1970; Vol. I, Part I, p 101.
10. Eliade, M. *The Forge and the Crucible: The Origins and Structure of Alchemy*; Harper and Row: New York, 1962; p 46.

Timeline—Hellenistic and Roman Eras

	600–300 B.C.	300–125 B.C.	125–50 B.C.	50 B.C.–225 A.D.	225–400 A.D.	400–575 A.D.	575–750 A.D.
Advances in Natural Philosophy	Beginnings of esoteric alchemy, c. 600	Bolos of Mendes (300–200), famous alchemist	Hermes Trismegistos, legendary alchemist	Posidonius writes of effect of moon on tides, 100 A.D.	Zosimus, earliest definitely known alchemist, c. 300	Museum at Alexandria closed, 415 Academy of Plato closed, 521	
Historical Landmarks	Conquests of Alexander; beginning of Hellenistic era, 335–325	Start of Roman conquest, 300	Rome becomes supreme in Mediterranean, 200–46		Constantine, first Christian emperor (300–337), moves capital of Roman Empire to Constantinople; beginning of Christian era Serapion sacked, 389	Rome sacked by Goths, 410 End of Western Roman Empire, 476 Nestorians move to Persia, 489 Triumph of Christianity, 491; Theodosius outlaws paganism	Arabs conquer Near East, North Africa, Spain, 640–720; end of dominance of Greek language and culture in southern and eastern Mediterranean

IV

Islamic Alchemy

THE MOSLEM EMPIRE was the largest the world had yet known. At its peak, about 750 A.D., it stretched from the Atlantic coast of Portugal to the western borders of China and included Spain, Portugal, all of North Africa, the Near East, and most of Central Asia. This enormous area contained all of the cultural and technological centers of the Middle East and the Mediterranean, except for Greece and Italy.

The Arab conquests of 632 to 750 resulted in enormous intellectual cross-fertilization. The world of Islam[1] was a huge economic and cultural free-trade area, and the greatly increased trade and travel between different regions of this vast empire stimulated the flow of information and knowledge everywhere. For the first time in almost a thousand years there was large-scale cultural and commercial exchange between Persia and Egypt. North African merchants visited India and China, and Arab traders traveled as far north as the Ukraine and even to Sweden. Between the years 750 and 1150, the Islamic world reached a level of prosperity and cultural attainment not matched in the West until the late years of the Renaissance.

---

[1]Islam is the collection of states or societies professing the religion of Mohammed. The term is analogous to Christendom. A Moslem is a believer in the religion of Mohammed; an Arab is one whose language is Arabic. Many Arabs, such as Copts and Maronites, are not Moslems and many Moslems, including Turks, Iranians, Indonesians, and Afghans, are not Arabs.

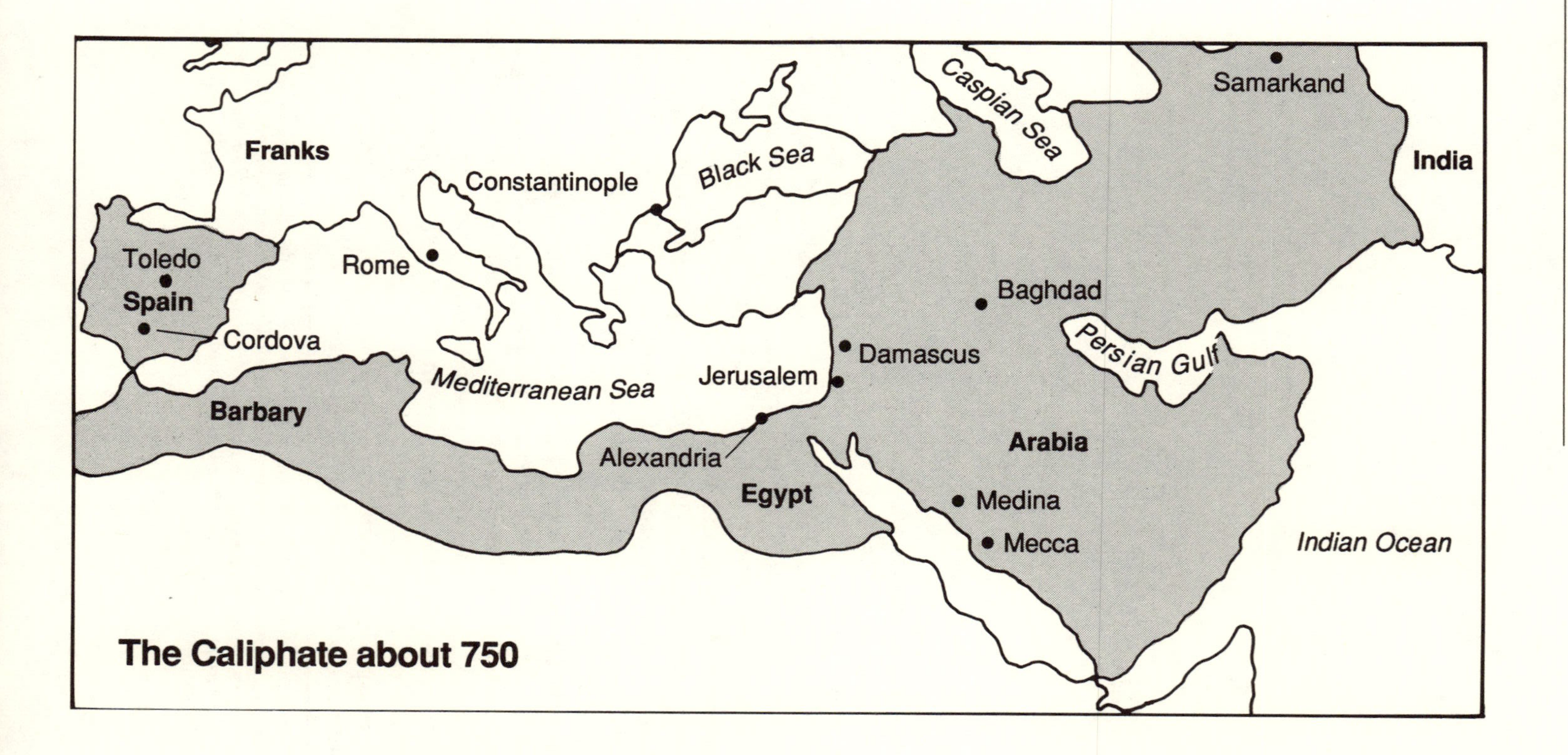

The Caliphate about 750

Islamic Scientific Activity

When the Arabs moved out of the desert, unlike other barbarian conquerors, they were well aware of the culture and accomplishments of their new subjects. They rapidly developed an interest in natural philosophy, and from about the middle of the seventh century through the next four or five hundred years, scientific activity was predominantly Islamic. Nevertheless, Christian and Jewish scientists were active in the Islamic territories. The Moslems were relatively tolerant because they considered Judaism and Christianity to be the precursors of Islam, and at first they permitted Christian, Jewish, and even pagan schools. (They were not at all eager to convert their subjects, because only non-Moslems paid taxes and massive conversions would have lowered tax revenues.) Later on, attitudes changed, and religious persecution was as prevalent in Islam as anywhere else.

Today, as Islamic manuscripts are translated and the translations made available, it is becoming apparent that Arabs, Moors[2], and other Moslems made important discoveries. In the Moslem area, there was an enormous increase in scientific and technological knowledge. Islamic scientists made the first attempts at organizing chemical knowledge by trying to classify reactions. They drew, for the first time, a distinction between alchemy and chemistry, the former term being reserved primarily for attempts at transmuting metals. They modified Greek theory, improved previously known processes, especially those of distillation, and worked with many more substances than the Greeks had known. For example, in Islam in the early medieval period, solutions containing nitric acid were being used, probably as a solvent and for medications (*1*).

Considerable advances were made in pharmaceuticals. Moslem physicians were far superior to both their ancient predecessors and their European contemporaries. For example, in 870, Abu Mansur used plaster of Paris casts to immobilize fractures. His pharmacopeia included 585 drugs, of which 75 were of mineral origin, 44 of animal origin, and 466 of vegetable origin. With so many different medications to deal with, the Arab apothecaries, who had originally been grocers and sellers of herbs and spices, became professional pharmacists. As early as Abu Mansur's time, some were specialists in compounding drugs, and others were merely retail distributors. The Moslem pharmacists introduced pills, powders, juleps, syrups, poultices, and court plasters. Most of their remedies were "simples," composed

[2]Moors are North African Moslems of mixed Arab and Berber descent. They subjugated Spain in 712.

of only one ingredient, and many of them are probably still in use in Africa, India, the Middle East, and South America.

The Arabs obtained their first knowledge of Hellenistic science from the Nestorian Christians who had translated Greek documents into Aramaic. Also, when they overran Syria, Egypt, Spain, and northern Africa, the Arabs acquired a great many Greek and Latin scrolls from monastery libraries, the private collections of the few wealthy pagan and Christian classicists, and the remnants of the library of the Museum at Alexandria. In 873 the caliphs instituted a program to translate these and all other Greek, Latin, and Aramaic philosophical and scientific works into Arabic. When they realized that the poetic language of the desert warriors, although superb for romances of love and valor, was not adequate for scientific and philosophical literature, they ordered their grammarians to reorganize and codify Arabic grammar. The result was the organized logical Arabic grammar that preceded the work of Richelieu's grammarians at the Académie Française by almost a thousand years. It was, and still is, a great success and remains in sharp contrast to the relatively anarchic state of English grammar. As a result of the Arab systematic program of translation, almost all the Greek literary, philosophical, and scientific works that had survived the onslaughts of time, wars, and riots were translated and became part of Moslem cultural resources, available to Moslem scholars as far east as Samarkand in central Asia and as far west as Cordoba in Spain.

Chinese Contributions

In its rapid initial expansion, the Arab caliphate reached the Chinese border, and, after some fighting, commercial and cultural relations were established. The Chinese and Arabs were now neighbors, and Chinese ideas, as well as Chinese silks, moved westward across the Old Silk Route to the Arabs and western Europeans. Among these Chinese ideas was a most important concept, that of the elixir of life.

Chinese alchemy had developed separately but along lines similar to those of western alchemy (2). Like the Greeks, the Chinese believed that there were only a few elements: fire, water, wood, metal, and earth. However, from the beginning, Chinese alchemy differed from western alchemy with respect to the purpose of transmutation of base metals into gold. The Hellenistic exoteric alchemists wished to make gold to become rich. Not so the Chinese alchemists. In China, wealth and status were based not on money but on land, birth, and royal favor. Its alchemists sought gold not for material advantage but for eternal life. Gold did not corrode, nor was it attacked by any reagent. Therefore, it was eternal, and ingesting gold or an elixir containing gold would convey immortality. A tract from the *Tsan T'ung Chi* (roughly translated, *Synergistic Herbal Medicines*) of Wei Po-Yang, a treatise on alchemy written about 144 A.D., says:

> Gold is incorruptible and therefore the most valuable of things. The men of Art, feeding on it attain longevity . . . hoary hair regains its blackness and new teeth grow where fallen ones used to be. If an old man, he will once more become a youth. If an old woman, she will regain her maidenhood. Such transformations make one immune from worldly miseries [3].

The desire to live forever goes back a long way into prehistory, so there are no records of the origins of the Chinese hope that eating a magical elixir would confer immortality or at least longevity. Perhaps the Chinese received the concept of the elixir through their contacts with India, where hallucinogenic mushrooms were used by Vedic priests. Such mushrooms contain psychotropic drugs that can pass through the body into the urine, and the Vedic hymns mention that those who drank the urine of a priest who had eaten sacred mushrooms believed themselves to be immortal. Or perhaps the idea of the elixir came originally from Mesopotamia. The Sumerians were in contact with their contemporaries in India, and the *Gilgamesh Epic* of the Sumerians, composed about 3000–2000 B.C., mentions an herb that confers immortality. Probably, however, the Chinese developed the concept of the elixir of life from their own traditions and viewpoints. Joseph Needham, in his monumental *Science and Civilization in China* (4), comments that the Chinese desire for eternal life resulted from their practical, worldly outlook. There was no concept of afterlife. Naturally, because there was only one life, the Chinese wanted it to be preserved, forever if possible.

The connection between gold and the elixir of life is also uncertain. In Egypt, Syria, and Mesopotamia, gold had been believed to be the purest of earthly things, the blood and semen of the gods. In India, too, gold was the symbol of purity. These ideas might have been common to the ancient Chinese as well. Or the Chinese may originally have confused gold with antimony sulfide or stannic sulfide; in some mineral modifications, they look like gold. Antimony sulfide, swallowed in small doses, can cure gastrointestinal disorders (*see* Chapter VII). Cures by something that looked like gold might well have given rise to the belief that soluble, or potable, gold was indeed the elixir of life.

Later on, however, elixirs were prepared from mercury, lead, copper, and silver as well as "gold." In any event, by perhaps 400 B.C., the Chinese believed firmly that potable gold was the elixir[3].

The idea of the elixir of life spread westward slowly, and it did not reach the Mediterranean world until after the Hellenistic–Roman period. However, by about 900 A.D. the Moslem alchemists had accepted the search for the elixir as one of their prime objectives, and from them the idea was passed along to the West.

Another essential contribution from China transmitted westward via the Arabs was paper. Without paper, printed books would not have been possible. Knowledge would have been restricted to the few who could afford costly books written by hand on parchment. In 751, at the battle of the Talas River, near Tashkent, the Arab soldiers captured some Chinese papermakers and then set up a paper mill at Samarkand. The technology spread swiftly through Islam and to the West, providing paper for the increased needs of both commerce and the newly emerging bureaucratic states for ledgers, account books, tax lists, journals, letters, and other records. The demand was so great that by 793 a paper mill had been set up in Baghdad. Paper mills were built in Spain between 1100 and 1150, and the first such mill in France was built in 1198.

Other Contributions

The Chinese contribution to Moslem science was minor, however, compared with that of the Greeks. The Moslems adopted both Greek natural philosophy and alchemy as modified by the Alexandrians, although some Moslem philosophers had doubts about the practicability of transmutation. They were also influenced by ideas and practices deriving from Mesopotamia and India, with which the Alexandrians had had little contact. Oddly enough, at the start of the Moslem era, the great center of Mesopotamian alchemy was no longer in Mesopotamia itself but in the town of Harran in what is now Turkey. The Sabians, the people of Harran, had retained their original Sumerian–Babylonian religion and customs throughout the thousand years that they were ruled by Parthians and Persians. They were tolerated by

[3]An interesting consequence of this belief is the current Chinese euphemism for suicide: to "drink gold." Needham (4) remarks that for centuries the Chinese quite uncritically ate all kinds of potions that were supposed to be potable gold. Many people, including at least three emperors and many high officials and members of the royal family as well as countless commoners, died from drinking them. The reaction to these deaths was the cynical but realistic comment that drinking gold was equivalent to committing suicide.

the Moslems and continued in their ancient ways until ultimately the town and its people were destroyed by the Mongols about 1300. The Sabians practiced the techniques of the Sumerians, especially in preparing perfumes by distillation and elixirs from animal and vegetable substances. (The Sumerians themselves, of course, had not prepared elixirs, which were undreamed of until some two thousand years after the end of the Sumerian period.) Sabian ideas and techniques were adopted and used by Jabir and al-Kindi, who will be discussed shortly.

Although initially, Islamic scientific theories were basically Greek, and their equipment Mesopotamian, the Moslems did make major contributions. Their philosophers were deeply interested in matter, motion, space, and time, and were well aware of the views of Democritos, which they modified significantly. To Democritos, the properties of matter were those of the constituent atoms. Copper was red, for example, because each copper atom was red. But to the great Islamic thinkers, the properties of matter resulted from the way atoms combined. Copper was red because of the way copper atoms were arranged in a piece of copper. (This belief was perhaps the first glimmering of the idea that observable properties depend on the arrangement of atoms. The conflict between the viewpoint of the Moslem atomists and that of Democritos was settled only in the mid-nineteenth century (*see* Chapter XIV). As a result, the Moslems were much concerned with the question of minima, the smallest particles that could still retain the recognizable properties of the substance (5). If a piece of salt were made smaller and smaller, at what point would it stop being salty? How large was a minimum? Abu Rasid (c. 1000) considered two atoms to be the minimum, while Abu al-Hudail (c. 950) thought six atoms would be the minimum. Some Arab philosophers, as reported by the medieval Jewish scholar Maimonides, believed that not only matter but space, time, and motion were discontinuous, a startlingly modern idea. These views were certainly known to William of Ockham, Nicholas of Cusa, Nicholas d'Autricourt, and other fourteenth century natural philosophers (*see* Chapter VI) and were passed along to Leonardo da Vinci and the sixteenth century thinkers.

Moslem Natural Philosophers

The Moslem natural philosophers were thoroughgoing materialists, very much interested in the results of experimentation. They stressed experimental verification of ideas, careful preparation of materials, and testing for purity. Because the huge Moslem free-trade area made many more naturally occurring materials available, the Arabs knew and used a great many more chemicals than the Alexandrian Greeks. Many of these were similar to other chemicals, and so they were forced to check the identity and test the purity of their materials. The physician al-Samarquandi (d. 1222) used more than six hundred chemical materials for which there were tests of purity and whose chemical and physical properties were known and listed (*6*). In fact, so many substances were known that the Islamic alchemists and physicians were forced to classify them, which they did on the basis of physical properties, such as fluidity and flammability. This was an important step toward the organization of chemistry as a science. Actually, this classification of chemical substances underlines the shift in emphasis from Aristotle's qualities to matter itself.

Jabir

The most famous of the Moslem alchemist–chemist–scientists were al-Kindi, al-Razi, ibn-Sina (or Avicenna, as he is known to the West), and Jabir ibn-Hayyan. Jabir is the earliest great name we know in Moslem alchemy, although perhaps the man himself never existed. He was supposedly born about 720 and died, or was perhaps murdered, about 815, at an astonishing age for that era. He was a member of the dissident Isma'ilia sect, which was deeply involved in esoteric alchemy. Its members were also political activists, and one of their branches was the famous secret society of Hashashins, or Assassins. These people were terrorists and political murderers who killed under the influence of hashish. The Hashashins were stamped out by the Mongols in the thirteenth century although the Isma'ilia sect still exists, its present leader being the Aga Khan.

Another of the offshoots of the Isma'ilia was the secret society of the Faithful Brethren or Brethren of Purity, which may have produced the books ascribed to Jabir. It is now quite well established, on the basis of internal evidence such as grammar, spelling, and reference to events that took place long after 815, the supposed year of his death, that most of these works were not written by Jabir, although some of the earlier ones may have been. The complete body of the works attributed to Jabir includes more than two thousand books, and of course no one man could have written that number, even if he did live to be ninety-five. Obviously, most of these books were written over a period of many years by different authors who attributed their writings to Jabir, either out of respect or as a method of remaining safely anonymous.

Although the scientific ideas of the Jabir corpus were basically those of Aristotle, there were some important differences. Contrary to Aristotle's explicit doctrine, in the Jabir works each substance was considered to contain all four of the qualities instead of just two. The mutually antagonistic qualities, such as hot and cold, did not annihilate each other but coexisted, like the Chinese yang and yin. Another subtle but very important change was in the concept of the qualities and the elements. To Aristotle, the qualities and elements were abstract and philosophical. We might even use the term hypothetical. To the Jabir writers, the qualities and elements were real and might actually be isolated. They believed that destructive distillation as practiced by the Hellenistic alchemists broke materials down into their qualities and elements. The smoke and tar resulting from pyrolysis were therefore elements, although obviously not in pure form. Purification, however, could be achieved by distilling and extracting the tars and condensates again and again, the required number of distillations and extractions corresponding to some magic number. Water, for example, was to be distilled seventy times. (There were many elements of numerology in the Jabir work, although the bulk of these writings was relatively free from allegory and mysticism.)

The Jabir writers accepted the Hellenistic modification of Aristotle's two exhalations, the one smoky and the other vaporous, but modified it still further. By now, the number of exhalations had been expanded considerably, and they were called variously "spirits" or "essences" or "principles" (the names seem interchangeable). The smoky exhalations were earthy because they were believed to contain small particles of earth being transformed into fire. In other words, they burned. Among these earthy spirits were sulfur, orpiment (arsenic sulfide), and oil. The vaporous exhalations were supposed to contain water particles being changed into air. Mercury, sal ammoniac (ammonium chloride), and water were vaporous exhalations. Minerals were composed of various combinations of the two volatile spirits and a nonvolatile body, an idea developed further by the Swiss medical alchemist Paracelsus some seven hundred years later. Metals were combinations of "philosophic" sulfur and mercury (another Hellenistic idea), ordinary sulfur and mercury being too impure to combine and form metals.

The Jabir writers prepared solutions containing nitric acid, although not in pure form, and described processes for making steel, ink, and glass; for dyeing cloth and hair; and for waterproofing materials. They knew that bronze, brass, and electrum (a naturally occurring alloy of gold and silver) were mixtures and did not list them as metals.

Al-Kindi

We have a bit more detailed knowledge on the next great Moslem natural philosopher, the somewhat inconsistent al-Kindi (c. 800–870). He was a mathematician who translated and commented upon the science and philosophy of the Greeks and attempted to apply geometrical methods to science and medicine. He was also a practical laboratory technician. He believed in astrology, but he was skeptical about the transmutation of metals into gold and in fact attacked the practices of the alchemists.

Al-Kindi wrote books on music, optics, logic, astronomy, mathematics, and meteorology, among others. His scientific books include titles such as *Treatise on Distillation of Aromatics, On Various Kinds of Philosophers' Stones, On Dyes, On the Futility of the Claim of Those Who Pretend the Making of Gold and Silver and Their Deceits,* and *Treatise on the Fraudulent Acts of the Alchemists.* He believed that alchemists were charlatans and that silver and gold could not be produced by human agency. Nevertheless, he himself was a charlatan and a faker. One of al-Kindi's most important works, *Book on the Chemistry of Perfumes and Distillations,* contains 107 recipes for preparations of aromatic oils, salves, and perfumes, but it also lists adulterants for costly drugs and methods of preparing fake perfumes. In one recipe he boasts that he can fool even professional apothecaries. Although al-Kindi's ideas were much more scientific and he knew many more materials and reactions, his equipment for distillations and sublimations was really not much better than that of the Babylonian perfumers of 1200 B.C.

Al-Razi

Al-Razi (866–925), known to the West as Rhazes, was more ethical and of a more philosophic character than al-Kindi. He was a deist, perhaps even an atheist, and certainly a heretic. He wrote that religions were the sole cause of wars and that the Scriptures were worthless. Of Persian birth, he studied in Baghdad, in Mesopotamia, where a series of enlightened caliphs had been patrons of literature, science, and philosophy and had set up research institutes specializing in the study of Hellenistic science. There were famous scholars and universities with libraries containing translations of Plato, Aristotle, Galen, and the other Greek philosophers and scientists. Although al-Razi wrote on chemistry, he was primarily a physician, considered to be second only to the great Avicenna (*see* next section) among Islamic doctors. He recognized that plague, smallpox, and consumption were infectious diseases, although this contradicted accepted theory. He was the first in the long and distinguished line of physicians of all creeds who virtually dominated alchemy and chemistry until the late eighteenth century.

Although he believed in transmutation, al-Razi was not interested in the mystical or allegorical aspects of alchemy. He was a clear-headed practical laboratory technician and observer who classified not only substances but apparatus and reactions. He wrote some thirty-three books on natural science, eleven on mathematics, and more than a hundred on medicine. In his most famous work, *The Book of the Secret of Secrets*, he attempted to debunk the belief that there were indeed secrets known only to the initiated.

Al-Razi classified chemicals by origin: animal, vegetable, mineral, or derived from other chemicals. Among the substances of animal origin were bones, milk, eggs, urine, and hair. Minerals were subdivided into spirits, fusible elements (which apparently comprised the metals), stones, boraxes, and salts. He mentioned the Arab discovery of the caustic alkalies (the alkali hydroxides). In what is possibly the first mention of acids as a class, he listed vinegar, sour milk, and lemon juice among the "sharp waters" useful for dissolving metals and various other substances (7).

He used the traditional hearth, bellows, crucible mold, and oven for smelting metals. His general instruments included the still, beakers, and receivers, and his procedures included distillation, sublimation, filtration, roasting, digestion, amalgamation, calcination, and decantation, most of which were known to the Sumerians. In general, his techniques were not much more advanced than those of the Alexandrian Greeks, but he gave detailed instructions in a logical manner, explaining at each step what he was doing and why he was doing it. He accepted the belief that substances were combinations of spirit and soul with inert material, which they activated, and his chemical reactions and processes were intended to change the relative proportions of spirit and soul by various purifications.

Avicenna

The best known of the eastern Islamic natural philosophers was ibn-Sina (980–1037), called Avicenna in the West. He was a poet, a scientist, a philosopher, and is also considered to be the greatest physician of medieval Islam. He wrote more than one hundred books, one of which was a medical encyclopedia, *The Canon of Medicine,* that was the definitive work in both the East and the West up to the time of Paracelsus in the sixteenth century and is probably still in use today in some underdeveloped regions. In it, he set up rules and procedures for the proper experimental study of the dosages and effects of drugs, and he attempted to relate the effects of the drugs to the proportions of their qualities. His procedures were inductive rather than deductive.

He did considerable work in geology—writing on minerals; describing conditions for the formation of stones, rocks, and mountains; and proposing a quite modern theory for the formation of fossils. His works were translated into Latin in about 1200 and were incorporated into the medieval Christian encyclopedias.

Avicenna's contribution to chemistry lay in his clear and logical exposition of ideas. He did not do much original work; the one new idea we can definitely attribute to him was the proposal that chemicals maintain their identity even in compounds. Aristotle, it will be remembered, had postulated that when substances react, each one loses its identity, forming a homogeneous product in which there is no trace of the original constituents. Avicenna calmly asserted that Aristotle had been wrong. Nevertheless, even though Avicenna's works were widely read and very influential, this particular suggestion had little effect. It remained for Angelus Sala in seventeenth century Germany to produce an acceptable demonstration that Aristotle had been wrong. (Even then it made no great stir.)

Avicenna was more than skeptical about the transmutation of base metals into gold—and in fact about alchemy in general:

> As to the claims of the alchemists, it must be clearly understood that it is not in their power to bring about any true change of species. They can, however, produce excellent imitations. . . . Yet, in these the essential nature remains unchanged. These properties which are perceived by the senses are probably not the differences that separate the metals into species [another direct contradiction of one of Aristotle's fundamental ideas], but rather accidents or consequences, the specific difference being unknown. And if a thing is unknown, how is it possible for anyone to endeavor to produce it or to destroy it? [*8*]

In this translated excerpt, it is clear Avicenna did not deny the validity of the Four Elements theory but he did not believe that human beings could perform transmutations. He also clearly distinguished between the alchemists and himself, here and in his other works also. He classified himself as either a physician or a chemist and reserved the term "alchemist" for those attempting to produce gold and silver.

Avicenna lived at the end of the three-hundred-year golden age of eastern Islamic science. By about 950, the religious and political unity of Islam had been fragmented. A body of Moslem dogma had appeared, and the initial Arab tolerance and skepticism had changed to suspicion, intolerance, and bigotry. Science and scientists were especially suspect.

Eastern Islam broke up into warring fragments where study, contemplation, and cultural interchange became difficult, if not impossible. Shortly after 1000, Mesopotamia and part of Persia were overrun by the Turks, a warlike people with contempt for learning and an intense conservatism in religion and philosophy. The coup de grâce came about 1225 when the Mongols brought death and devastation as far west as Venetian territory. They burned Baghdad, exterminated the Sabians at Harran, and looted and sacked all through Persia and Mesopotamia.

Alchemical Science in Western Islam

By then, however, scholars had carried Moslem science westward to Sicily and Spain and brought with them the translations of Hellenistic works that had been circulating freely in the eastern Islamic world. The works of Jabir and al-Razi were known in Spain before 1000. From about 950 on, there seems to have been a flourishing group of alchemists and chemists in Spain, although their names are not known with any certainty. In any event, there was little time left for them. The Christian reconquest gained momentum about the year 1000, and in 1085 Toledo, the ancient Spanish capital, was recaptured from the Moors, who were thereafter on the defensive and were steadily pushed back. Finally they had to seek military help from the fanatical Moslem Berbers of North Africa. Southern Spain remained Moorish, but at the price of its freedom. The Spanish Moors became the subjects of a narrow theocracy that persecuted alike Christians, Jews, dissident Moslems, and any unorthodox thinkers.

The two most influential Moorish works are the treatises *De Aluminibus et Salibus* ("Concerning Alums and Salts") and

Alchemica de Anima ("About the Alchemical Spirit"). The first of these was probably written in the late eleventh or early twelfth century and was translated into Latin by Gerard of Cremona (1114–1187). *De Aluminibus* is a concise set of directions and formulas for making spirits (distillates), salts, metals, and stones. Among other things, it describes the production of corrosive sublimate (*9*).

Corrosive sublimate is mercuric chloride, a deadly poison that the pharmacist calls bichloride of mercury. It is an antiseptic and also a chlorinating agent, for on being heated it breaks down into chlorine and mercurous chloride, which was a widely used medication called calomel. Mercuric chloride was therefore an important chemical. It was prepared either by heating mercury, alum, and ammonium chloride or by heating mercury with salt and vitriol (an impure ferrous sulfate usually found in pyrite deposits). This successful synthesis of corrosive sublimate attracted attention to the synthesis of other materials that later became part of the arsenal of the alchemist and the physician. In a very important modification of the process, sal ammoniac and vitriol were heated together without the mercury. The product was the alchemists' "spirit of salt," our hydrochloric acid.

De Anima was written about 1200 but credited at that time to Avicenna. It is another in the large number of works—going all the way back to the biblical passages now attributed to Deutero-Isaiah—that the author, for some reason, pretended to have only discovered. The unknown author of *De Anima* is now referred to as pseudo-Avicenna. The book is a collection of summaries of the opinions of authorities on the theory of alchemy and the reasons for the procedures, and is generally considered to be a compilation by the various authors of the Jabir corpus.

Another important Moorish text, *The Sage's Step*, was attributed to the astronomer al-Majritie, who died in 1007, but it was probably written after his death. It contains a series of recipes that show a quantitative attitude and considerable laboratory experience on the part of the author:

> I took natural quivery mercury, free from impurity, and placed it in a glass vessel shaped like an egg. This I put inside another vessel like a cooking pot and set the whole apparatus over an extremely gentle fire. The outer pot was then in such a degree of heat that I could bear my hand upon it. I heated the apparatus day and night for four days, after which I opened it. I found that the mercury . . . had been completely converted into a red powder, soft to the touch, the weight remaining as it was originally [*10*].

In this passage the author has described the use of glass apparatus, the purification of reagents, the control of temperature, and reliance on the measurement of mass. Surprisingly, in the work of a laboratory chemist of such sophistication, there is an important error. If complete conversion had occurred, the red powder would have weighed some eight percent more than the mercury. Some seven hundred years later, Antoine Lavoisier used the same reaction to help establish the modern view of combustion.

During the twelfth and thirteenth centuries, a number of Moorish alchemists wrote books, but to escape persecution by the orthodox Moslems and the religious authorities, they claimed merely to have discovered the lost works by previous authors. Inasmuch as the same sort of precautionary pseudonymity was being practiced in Christian Europe for much the same reasons, considerable confusion has ensued. A number of Latin writings by Geber were first attributed to Jabir, but they must have been written much later than Jabir's time, because they contain information that the original Jabir authors could not possibly have had. Many materials and processes are described that were unknown to the eastern Arabs. Also, no Arabic works by Geber have ever been found, so it is now believed that Geber's works were originally written in Latin, not Arabic, and that the author only claimed to have translated them. He was much safer that way.

Although Moorish alchemy continued on a reduced scale (and is still continuing [*11*]), after the end of the eleventh century the major new developments in chemistry were in Christian Europe. Nevertheless, the average Moorish chemist possessed knowledge and equipment far beyond that of the Alexandrian Greeks.

Arab science had a brilliant beginning, but it was stifled by wars and invasions that broke up the universities and destroyed the major centers of culture, by religious and political repression, and by the lack of inexpensive and readily available books.

References

1. Holmyard, E. J. *Alchemy*; Penguin: Harmondsworth, Middlesex, England, 1957; p 79.
2. Davis, T. L. *Isis* **1938,** *28,* 73.
3. Lu-Chi', W.; Davis, T. L. *Isis* **1932,** *8,* 240–241.

4. Needham, J. *Science and Civilization in China;* Cambridge University Press: Cambridge, England, 1974; Vol. 5, Part 2, pp 13–14, 71, et seq.
5. Levey, M. "Studies in the Development of Atomic Theory," *Chymia* **1961,** *7,* 40–56.
6. Levey, M.; Al-Khaledy, N. *Al-Samarquandi;* University of Pennsylvania Press: Philadelphia, 1967; p 41.
7. Stapleton, H. E.; Azo, R. F.; Husain, M. H. *Mem. Roy. Asiatic Soc. Bengal* **1927,** *8,* 392.
8. Holmyard, E. J.; Mandevelle, D. C. *Avicennae De Congelatione et Conglutinatione Lapidum;* Paris, 1927; pp 41–42, quoted by Leicester, H. M. *Historical Background to Chemistry;* Dover: New York, 1971; p 70.
9. Multhauf, R. *Origins of Chemistry;* Franklin Watts: New York, 1967; pp 162–163.
10. Holmyard, E. J. *Isis* **1924,** *6,* 293–305.
11. Holmyard, E. J. *Alchemy,* op. cit., p 101.

Timeline—Islamic Alchemy

	600–700 A.D.	700–800 A.D.	800–900 A.D.	900–1000 A.D.	1000–1100 A.D.	1100–1200 A.D.	1200–1300 A.D.
Technological and Scientific Advances		Chinese papermakers seized by Arabs Arabs use Arabic numerals, 760 Arabian translation of Euclid's "Elements," 774 Jabir lived, 720–815	Abu Mansur born, 870; plaster of Paris bandages used al Razi lived, 866–925	Avicenna lived, 908–1037	*De Aluminabus* written	Physician al-Samarquandi used 600 chemical substances; *De Anima* written	
Historical Landmarks		Arabs extend conquests to central Asia, 632–750 Charlemagne crowned Holy Roman Emperor, 800	Caliph Haroun al Raschid rules, 789–809		Rise of Turks in Asia Minor, 1042; Turks subdue Syria, Palestine, 1075 Spaniards retake Toledo, 1085 Start of first Crusade, 1096		Start of Mongol invasions, 1218 Mongols destroy Sabians, 1225 End of last Crusade, 1291

V

Medieval and Renaissance European Artisans

IN EARLY MEDIEVAL EUROPE (approximately 700–900) about the only chemical activity was that of craftsmen decorating churches, forging weapons, dyeing fabrics, boiling soap, and making glass. Alchemy was largely confined to the southern and eastern Mediterranean, the Arab world. It had largely disappeared in the West and was reintroduced during the Crusades in the eleventh to thirteenth centuries.

The Roman Empire had been essentially a Mediterranean civilization with the greatest part of its wealth, commerce, and culture located in the East, in Egypt, Syria, and Asia Minor. To Imperial Rome, western Europe was primarily a source of raw materials and agricultural products, with some notable exceptions. Its eastern possessions were much more important than the western provinces, and Rome kept its best troops on the eastern border, facing the Parthians and their successors, the Persians, not only because the eastern lands were more valuable but also because the Persians were far more formidable militarily than the Germans. The Romans also shifted the seat of government eastward, to Constantinople. In the long run, this strategy paid off. After the Empire split in two in 395, the eastern, or Byzantine, half survived for more than a thousand years, until

1453. In the short run, however, the West, weakened militarily and financially, was overrun by a relatively small number of German tribesmen.

At the end of the Germanic invasions (about the middle of the sixth century), the former western empire was divided into a group of petty Germanic kingdoms, those of the Ostrogoths, Franks, Visigoths, Burgundians, and Lombards. Economic, social, and psychological conditions had changed completely. The Roman Empire had been a multinational free-trade area with large-scale commerce and industry, international trade, and an economy based on money. The Roman state had a well-organized bureaucracy and a tradition of secular education in practical matters for aspiring merchants and civil servants and classical education for the aristocrats. In the post-Roman Germanic kingdoms, however, power was in the hands of an uneducated illiterate military aristocracy. Commerce and industry were on a much smaller scale. Hard money was scarce, and there was a growing trend toward barter, especially in rural areas.

Even before the Germanic invasions, as the Empire became Christian, education had begun to shift toward clerical concerns. Secular schools began to die out except in the commercial centers of Italy and the East, where merchants and clerks were given the necessary business training. Classical pagan learning decayed almost completely except, paradoxically, in the Church, where the early Church fathers were staunch believers in the merits of the education they themselves had received. Saint Jerome, a classically educated philologist, translated the Bible into Latin, producing the Vulgate edition. Saint Augustine, too, was well educated and in fact had been a teacher of rhetoric. Augustine recommended the study of grammar and rhetoric and of compendia of liberal arts, such as philosophy and literature. (These were summaries, not thorough studies of philosophy and literature.) Augustine and Jerome supported classical education partly because of tradition and partly because it inculcated virtues, but mostly because the Church needed educated men. In fact, the Church, based as it was on the written Gospels, was threatened by the loss of literacy.

Nevertheless, in spite of the efforts of Jerome and Augustine, educational conditions in the West worsened until Flavius Cassiodorus (490–585), managed to revive some part of the classical heritage. Cassiodorus was a wealthy classically educated Roman official in the service of Theodoric (c. 460–526), the Ostrogothic king of Italy. At first he attempted to found a school in the old tradition, but the effort failed. Then he made use of the monastic movement. On his large estate, he established two monasteries and introduced a program of education for the monks. This course included not only Scripture, theology, and Church history, but also a prerequisite curriculum of the liberal arts. The monks exposed to this education were literate and had at least a slight exposure to classical subjects, obtained by studying specially written texts. Cassiodorus introduced the practice of copying classical manuscripts as a worthy monastic occupation, a practice that spread to other monasteries. As the medieval era continued and the number of monasteries increased, the monks began to send copies of their manuscripts to other monasteries. The love of learning took over and some monasteries became centers of copying, the early equivalent of publishing houses. As the manuscripts accumulated over the centuries, the monastery collections eventually evolved into rich libraries. It was largely these libraries that preserved the classical heritage in western Europe and ultimately transmitted it to the Renaissance humanists. The world owes much to Flavius Cassiodorus, and it is pleasant to note that he lived to the venerable age of almost a hundred years.

The Church needed and produced educated men. Illiteracy might be tolerated in rural priests, who cared for the spiritual needs of peasants and serfs, but not among the functionaries of the great churches and cathedrals, which had more complex problems. Church and cathedral schools were founded all over western Europe to train priests and Church administrators. Although some graduates went into the service of the new kingdoms, where they were desperately needed, most went directly into Church service.

The Church, however, needed more than clerks. It needed masons to build churches and painters and stone carvers to decorate them. It needed leather workers and copyists for the holy books, winemakers for the sacramental wines, and coopers for barrels to store them. These and other artisans had to be trained, and training required both instruction manuals and literate teachers. Moreover, the attitude toward manual work had to change.

In Greek and Roman times, manual labor was considered degrading, but in early medieval Europe, even gentlemen of the lesser nobility might do agricultural labor. Monks, even those of aristocratic origin, worked with their hands. Perhaps this more tolerant attitude toward work resulted from the manpower shortage that forced the development of labor-saving devices and the technological improvements that by 1500 brought the

West to parity with the East. Most likely, however, it stemmed from Christianity's origin as a religion of proletarians and slaves, with the fisherman Peter and the carpenter Saint Joseph and other laborers as major figures.

Throughout the Middle Ages, during wars and invasions and enormous changes in European society, artisans of various kinds were working quietly at what later became industrial chemistry. They made soap, glass, metals, lime, plaster, lye, and dyes. They kept their methods secret and passed them down for generations by word of mouth, from father to son or master to apprentice. If for some reason, the procedures were written down, they were either in code or in allegorical style, so that no stranger could learn the secrets and go into competition.

Chemical craft practices were therefore very conservative; workers did what their remote ancestors had learned to do. For example, with most dyes, fabric had to be mordanted, that is, treated with chemicals to fix the dye to the cloth. With indigo a mordant was not needed. Nevertheless, throughout the entire medieval era dyeing with indigo was automatically preceded by the expensive, time-consuming, and unnecessary mordanting process. With such rigid attitudes, progress was slow, and rarely was there contact across craft lines. The soap boiler neither knew nor cared what the lime burner was doing, although they were both working with chemicals. Still, these artisans were responsible for some progress. Operating only by trial and error, they usually got nowhere in attempts to modify or improve processes. But over the course of the centuries, they did develop some new methods, new equipment, and useful new chemicals.

Metallurgical processes were constantly being improved, especially those involved in making armaments. There was also a large increase in the production of soap. Soap was primarily produced, not for use in personal hygiene but for the cloth industry, centered in medieval Flanders. Wool is naturally greasy, and the grease must be removed before weaving and dyeing; soap and alkali were needed in quantity. Alkali was also used in making glass for church windows, flasks, and drinking vessels. Major areas of glass production were Italy, Belgium, and the Rhineland, where alkali and fine sand were available.

Byzantine Contributions

In the East, the Byzantine Empire was on a much higher cultural, material, and intellectual level than the contemporary western kingdoms, but it was not a source of new scientific and technological knowledge. The Byzantine state was a highly organized and controlled bureaucracy where independent thinking was not encouraged, especially along philosophical lines. In general, the Byzantines were not innovators. They developed materials and methods for decorative arts to a high level, by improving existing materials and processes rather than by inventing them.

Greek Fire

There was, however, one supremely important exception, the famous Greek fire, invented by Kallinakos, a Hellenized Syrian from one of the great centers of Hellenistic alchemy. Greek fire was a mixture of petroleum or naphtha with sulfur and resins to thicken it to a jelly that would neither evaporate nor burn too quickly. Forced through a siphon and ignited by a flame burning at the tip, the spray of flaming jelly burned everything it hit and adhered to. The siphon was the prototype of the modern flame thrower, and Greek fire itself was the original napalm, which was used so extensively in World War II and later in Vietnam.

In 672 and again in 717, Greek fire saved the key strategic fortress Constantinople from being taken by the Arabs and thereby kept eastern Europe from being overrun. After these two great Arab assaults failed, Europe did not have to face another Moslem onslaught until the fifteenth century, when the Ottoman Turks took Constantinople and finally ended the Roman Empire. Greek fire was, therefore, a weapon of enormous historical impact.

When Greek fire was used against the invading Arab fleets in 672, the wooden ships burned, and the crews were burned or

Greek fire, the first known chemical warfare agent, being used in a naval battle. Taken from a tenth century Byzantine manuscript. (Reproduced with permisssion from reference 1.)

drowned. The panic it induced was as effective as the fire itself. Greek fire was also tried in land warfare, but there it was relatively ineffective, although it was useful in defending besieged cities. There are always many witnesses to a battle, so the secret of its composition could not be kept, and within a few years the Arabs too were using Greek fire. But by that time, their momentum and political unity had been broken, and they were more interested in retaining their conquests than in making new ones.

Collections of Recipes and Procedures

The Byzantines were much concerned with the ornamentation and beautification of churches and palaces. Their churchmen and state officials wore splendid robes and jewels as a matter of public policy. They compiled collections of recipes and procedures for enameling and painting and for the construction of mosaics and other artwork. Some of these were original with the Byzantines, although much had been copied from earlier texts. Those in turn had been copied from still others, going all the way back to the Assyrians (2), who themselves admitted to getting much of their material from the Sumerians (3).

The earliest of these manuscripts that we have is an eighth century copy of an Alexandrian work of about 600 A.D., the *Compositiones ad Tingenda* ("Recipes for Coloring") (4). It deals with making and coloring the glass used in mosaics, gilding, and dyeing leather. For the first time that we know of, the word vitriol is used. The text also contains the first known mention of the preparation of cinnabar (mercuric sulfide) by heating mercury and sulfur together. Both vitriol and cinnabar had, of course, been known for many centuries before the *Compositiones* was compiled. Another such collection of recipes is the *Mappae Clavicula* ("Little Key to Painting"), reportedly compiled in southern Italy or Sicily about 800, although the earliest extant copy dates from the tenth century. This manuscript gives possibly the first recorded mention of alcohol. Among other things, it reports that wine, when distilled, gives off a liquid that, upon being ignited, burns without setting anything else on fire. In other words, the distillate is a very dilute solution of alcohol in water.

Still another collection of recipes is *De Coloribus et Artibus Romanorum* ("On the Colors and Arts of the Romans"), which is from the tenth century (4, 5). It was a craftsman's manual for making enamels and artificial jewelry. Many of the recipes in these three books are almost identical with one another and with the Leiden papyri and were probably copied from a common source (*5*). Hawthorne and Smith (*2*) believe that the works ultimately can be traced to ancient Egypt and Mesopotamia.

Technological Advances

From about the year 1000, the pace of technological advance quickened. There was prosperity, and prosperity created new needs, new expectations, and new demands. In response, new materials were bound to be developed.

Glass

The most important new material came quite early in the form of hard, clear, strong glass that melted only at high temperatures. This glass was not only of fundamental importance to chemistry, but also it was one of humanity's most important technological developments.

During the period 1000–1100, Italian technicians were improving glass by trial and error. They tried different reagent mixes, purified their starting materials, used higher temperatures, and produced better colors and clearer, stronger glass. Shortly before 1100, in Salerno, they developed a hard, tough glass that did not melt until temperatures were relatively high. Its strength made it of practical use for bottles and windows, especially cathedral windows, and its high melting point made it useful for chemical apparatus. Almost immediately, glass flasks, distillation apparatus, and all kinds of vessels were used in chemical operations. Robert Grosseteste (1168–1253), Bishop of Lincoln and teacher at Oxford, mentions optical experiments done with a glass urinary flask; apparently such flasks were common enough to be used for chamber pots. Glass is heat-resistant, nonconducting, transparent, acid-resistant, and, above all, plastic when hot. No matter what exotic shape is needed, it can be made of glass. Moreover, pieces of glass can be melted together, to become sealed without cement. It is difficult to imagine modern chemistry without glass apparatus.

Two achievements with far-reaching consequences resulted rapidly from the new improved glass: Spectacles were invented, and alcohol was isolated. By all odds, spectacles were the most important result of the new and improved types of glass manufactured in the West. They were invented shortly before 1286, in northern Italy, around Pisa (*6, 7*), and they were rapidly adopted everywhere. Along with printing and the wheel, eyeglasses are one of our most beneficial material inventions. Not

The first known example of spectacles: a fresco painting of Cardinal Ugo of Provenzano, by Tomasso da Modena, 1352. (Reproduced with permission from Fratelli Alinari, Florence, Italy.)

only would millions of nearsighted, farsighted, and astigmatic people be illiterate without glasses or have insufficient vision for a skilled trade, but even people with normal vision often lose ability to focus after the age of forty. Eyeglasses just about double the intellectual life span of the average individual.

Alcohol

Alcoholic solutions had, of course, long been known. The word alcohol is of Arabic extraction (*al-kuhul*). The Arabs had distilled wines and perhaps beers with relatively inefficient cooling systems and obtained dilute solutions that contained a little alcohol in a lot of water. The alcohol content was so low that this *aqua ardens* or *aqua flamens*, as reported in the *Mappae Clavicula*, burned without producing heat. But as the new harder and stronger glass became available, the condensing columns could be made longer and narrower than the age-old clay columns, and the greater ratio of cooling surface to volume improved the efficiency of condensation and increased the alcohol content of the distillates. Shortly after 1167, alcohol itself was isolated, most probably at Salerno.

Alcohol had enormous impact on chemistry, medicine, commerce, and even theology. To the esoteric alchemists and the clerical natural philosophers, distillation of liquids and sublimation of solids were akin to a religious experience, and the isolation of alcoholic solutions must have been electrifying. Just as the True Faith plucked the soul from the impure body, so, by fire, distillation produced a clear pure liquid from a murky brown brew. Moreover, to the astonishment of all who observed, the water produced from such a distillation actually burned with a blue, gemlike flame, although everyone knew that the nature of water was to extinguish fire, not to burn. Finally, the water, when imbibed, produced a very pleasing intoxication. The fascinating alcoholic distillates were named *aqua vitae*, meaning "water of life," and the name still survives in the Swedish word *aquavit*, the English and Scottish word *whiskey*, the French *eau-de-vie*, and the Slavic word *vodka*, all of which mean or come from roots that mean "water of life."

Monks had long produced wines for sacramental purposes, and they were among the first distillers in the West. When they first obtained alcohol, they thought it was the quintessence from which the heavens are made, and they studied its properties intensively. By 1288 the study had perhaps gone a bit too far, because the Dominicans were forbidden to distill alcohol, possibly because of its connection with alchemy. Nevertheless, other monks continued to produce alcoholic beverages such as chartreuse and benedictine.

To the artisan, the alchemist, and the apothecary, alcohol was important mostly because of its solvent properties. The alcohol molecule can be considered a modified water molecule. It not only dissolves many salts and other substances soluble in water,

A Renaissance depiction of a distillery furnace showing flasks, columns, and receivers. (Reproduced with permission from the Bettmann Archive.)

but it also dissolves many organic materials insoluble in water, such as oils, waxes, lacquers, and perfumes. With alcohol as the solvent, an entire new set of useful solutions was now available to the chemical worker.

Mineral Acids

Another important new development was the discovery early in the thirteenth century of solutions of the mineral acids, that is, of sulfuric, nitric, and hydrochloric acids and *aqua regia,* the mixture of nitric and hydrochloric acids that dissolves gold itself (*5*). Although the pure acids were not isolated until hundreds of years later, the dilute and rather impure solutions of these acids were immediately extremely useful because of their reactivity. Mineral acid solutions can dissolve all metals and most ores either at room temperature or in the water bath, a device for moderate heating, somewhat like a double boiler. Consequently, with these solutions, the alchemist, the metallurgist, the jeweler, the apothecary, and the physician no longer needed enormous furnaces in special workshops. Glass vessels at workbenches in shops or homes were sufficient. Entirely new classes of room-temperature reactions were now possible, and there was an enormous increase in the number of people who could do laboratory work. Because the rate of new developments is generally proportional to the number of workers in a field, the discovery of mineral acids ultimately resulted in a greatly accelerated rate of progress in chemical technology.

The immediate application of mineral acid solutions was in the analysis of precious metals. Europe had been starved for nonferrous metals from Roman times until 968, when copper was found in the Harz Mountains. Then in 1136, silver was discovered in Saxony, and there was a great rush of silver prospectors, like the California Gold Rush of 1848–1849. After the first high-grade lodes had been exploited, with each new discovery some sort of on-the-spot assay was needed to determine if the metal content of the ore warranted the expense of mining. By the fourteenth century, there was an assayer at or near each mine. The new solutions of nitric acid and aqua regia enabled these assayers to make quick, and largely correct, decisions.

Nitric acid and aqua regia were used for assaying more than ores, however. Late medieval Europe had large quantities of coins in circulation, some of dubious value. (Monarchs had developed the habit of debasing their own currency, making a quick profit by replacing the gold or silver with copper.) To the merchant who might be paid in *bezants* from Constantinople, *maravedies* from Spain, *zechins* from Venice, *florins* from Florence, *ducats* from Milan, and *livres* from France, the gold and silver content of the coins was of vital importance. Yet, each coin had a different nominal weight of silver or gold, and many were counterfeit or at least debased. The world of commerce had an acute need for quick and accurate methods of analyzing the gold and silver content of coins. The new acids were immediately put to use. If a drop of dilute nitric acid placed on a silver coin turned green, the silver contained copper. If a gold piece reacted with nitric acid, it failed the acid test.

Trade and the Chemical Industry

By the late medieval period, agricultural and industrial production and the volume and complexity of trade had increased to levels that were orders of magnitude greater than that of the Roman Empire. Each year during the period from 1350 to 1450, the town of Bordeaux shipped between one million and two million gallons of wine to England. In 1329 the Florentine house of Acciajoli shipped almost one million bushels of wheat from southern Italy to Genoa, Florence, and Venice. Banking and finance became international. In 1257, Italian merchants from Lucca bought Chinese silks in Genoa, promising to pay for them at the Fairs of Champagne, near Paris, the payment being handled by a bank in Piacenza, Italy. This level of economic activity quite contradicts the idea that the later Middle Ages were primitive and backward. About 1350 the Bardi and the Peruzzi banks of Florence loaned King Edward III of England approximately fifteen million dollars in gold to finance his war against France. He eventually defaulted, and the banks failed and left to Cosimo de Medici a virtual monopoly of Florentine banking, which the Medici used to gain political control of the city.

With such a lively international trade, the need for chemicals expanded and chemical processes were now performed on a relatively large scale. The cloth industry of Flanders and Italy and also that of England after the thirteenth century had grown so much that dyed and finished cloth was the chief manufactured export to the East, being traded as far as China, over the Old Silk Route.

Alum for Dyes

To fix dyes to the great quantities of cloth that were being produced and exported, enormous quantities of alum were imported from Egypt and North Africa. Then, in 1275, Genoa leased from the Byzantine emperor the alum works at Phocea, in modern Turkey, and expanded them. For two centuries, Phocea was the most important source of alum for the West, although there were other sources. The Genoese shipped about 8000 tons of alum from Phocea annually from 1278 until 1455, sending it through the Strait of Gibraltar directly to Flanders. These shipments, being bulky and heavy, required large ships for transport, and the Genoese became famous for the size of their ships. They also used their large vessels for transporting grain from Africa to England and on other long-distance voyages[1].

In 1455, however, the Turks captured Phocea, and thereby triggered an alum famine in Europe until a large deposit of alum stone was discovered at Tolfa, north of Rome. From 1465 to 1470, about 1500 tons per year was mined at Tolfa, providing much of the revenue needed for the rebuilding of Renaissance Rome. To safeguard this revenue, Julius II, the della Rovere Pope, had to

[1]The Genoese experience with long sea voyages in large ships was later put to good use by those famous Genoese seamen, Christopher Columbus and John and Sebastian Cabot.

pacify the area and bring the feudal barons under control[2]. The financial and political costs of the papal alum monopoly were so great and alum prices rose so high that by 1500 strenuous efforts were being made all over Europe, usually supervised by chemists, to produce alum from native clays containing aluminum sulfate. These were generally successful and soon each European state had its own sources of alum and its own cloth industry, although the Tolfa mine is still operating.

Wood Substitutes

With the increase of population, the expansion of farmland, and the resultant clearing of forests, wood became scarce just at the time that the demand for building and ship construction increased. After 1200, a search began for wood substitutes for construction, heating, and smelting. Already by 1190 in Liege, Belgium, iron smelters were trying unsuccessfully to replace scarce charcoal with coal. (This effort was a failure, but the search for a substitute for metallurgical charcoal continued for another four hundred years until Abraham Darby developed metallurgical coke in seventeenth century England.) For heating, coal was used on a large scale. Coals from Newcastle were shipped in bulk to London, and, by 1273, coal smoke was such a nuisance in London that there was a petition for the king to outlaw the burning of coal, the first known complaint about air pollution. By 1380 the combination of coal smoke and ordinary fog was producing the famous London fogs. (They disappeared rapidly in the 1950s and 1960s, when oil and gas were substituted for coal and fireplaces were regulated rigorously.)

Metal production increased enormously, and there was also a great qualitative improvement in metallurgy. At the end of the twelfth century, when Richard the Lion-Hearted went on Crusade, his cavalry was equipped with fifty thousand iron horseshoes. For production on such a scale, furnaces had to be much larger and more efficient. By 1200–1300, in the Rhineland, the old Roman furnaces for iron smelting had developed into large furnaces that the Germans called *Stückoffen*, the direct ancestor of the blast furnace. These were between 10 and 15 feet high and could produce up to a ton of iron at a single smelting.

Metal fabrication improved too. Church bells were always in great demand because, except for the position of the sun, the sound of church bells was the only way for the farmer or the

[2]The story of his military effort, with the army led by the Pope in person, has been told many times in novels and in biographies of Michelangelo, Macchiavelli, and Leonardo da Vinci.

villager to know the time. In the fifth and sixth centuries, bells had been made of sheet iron wrapped around a mold. By 700, bronze bells were being cast, and by 1250 iron bells were being used. Lübeck on the Baltic had a bell founders' street; apparently, here, as well as in other towns, bell-casting was a recognized and honored craft, and demand was so great that many specialists could practice their trade in the same place at the same time.

New Needs and Desires

For the new larger scale processes of the late Middle Ages, a new type of industrial technician was needed, and for him a new type of manual emerged, as exemplified by *De Diversis Artibus* ("On Divers Arts"), also known as *Diversarum Artium Schedula* ("Pages on Various Arts"). This was not just a collection of recipes copied from previous books but original work by an experienced craftsman, probably Roger of Helmarshausen, a north German Benedictine monk who lived sometime between the tenth and twelfth centuries. (Hawthorne and Smith, in their translation [2], state that the work dates between 1122 and 1123.) This was a specialist's handbook, a how-to book, the first of many. There were sections on the handling of the tools of the trade, for example. Yet it was not just a set of instructions to be learned by rote. It contained discussions of the basic metal-refining processes and clear laboratory directions for analyzing gold by cupellation and for separating it from silver, using several different new methods. The *De Artibus* is primarily concerned with the preparation and application of pigments in painting, manufacturing glass and imitation gems, and analyzing gold and silver. Its existence testifies to a prosperous western Europe with large numbers of technicians engaged in decorating cathedrals, churches, and palaces and in producing both real and imitation jewelry for the aristocracy and the rising commercial classes.

The great increase in wealth, as demonstrated by the large number of cathedrals, each of which required enormous capital expenditure, created new desires and needs. Medieval artisans made no contribution to chemical theory, but their increased production of commodities led to a capital surplus, and their development of new materials both satisfied and created needs. The end result was the accumulation of information, capital, and the material resources necessary for the later development of chemistry.

References

1. Singer, C., Eds.; *History of Technology*; Oxford University Press: Oxford, England, 1957, Vol. II. After Goodrich, H. B. *World Petroleum*; Vol. 10, Palmer Publications: New York, 1939; p 35.
2. Hawthorne, J. G.; Smith, C. S. *Theophilus on Divers Arts*; Dover: New York, 1970. Reprint of the 1963 University of Chicago edition, pp xix–xxx.
3. Thompson, R. C. *On the Chemistry of the Ancient Assyrians*; Clarendon Press: London, 1925. *A Dictionary of Assyrian Chemistry and Geology*; Clarendon Press: Oxford, England, 1936.
4. Leicester, H. M. *The Historical Background of Chemistry*; Dover: New York, 1971. Reprint of the edition by John Wiley and Sons: New York, 1956; pp 75–76.
5. Multhauf, R. P., op. cit., pp 153–157.
6. Rosen, E. "The Invention of Eye-Glasses," *J. Hist. Med.* **1956,** *11,* 13–46,183–218.
7. Angus-Butterworth, L. M. In *History of Technology*; Singer, C. et al., Eds.; Oxford University Press: London and New York, 1957; Vol. III, Chapter 9, Section VI; p 230.

Timeline—European Artisans

	500–650 A.D.	650–800 A.D.	800–950 A.D.	950–1100 A.D.	1100–1250 A.D.	1250–1300 A.D.	1300–1450 A.D.
Technological and Scientific Advances	Cassiodorus lived, c. 490–585	Kallinikos invents Greek fire, 671 *Mappa Claviculae*, c. 800	Medical school at Salerno, c. 900 *De Colorbus et Artibus Romanorum*		Alcohol prepared, 1170; used medicinally, 1200 Glass windows in private homes in England, 1180 Londoners petition to outlaw coal burning, 1243 Roger Bacon mentions gunpowder, 1249	Spectacles invented, c. 1288	
Historical Landmarks		Charlemagne crowned Holy Roman Emperor, 800		Norman conquest of England, 1066			Turks take Constantinople, 1453

VI

Medieval and Renaissance Alchemists and Natural Philosophers

UNTIL ABOUT 1000 A.D. in western Europe, there was almost no alchemy and very little scientific activity. Then, almost abruptly, a great economic and social change got under way. The population began to increase. An agricultural surplus accumulated, sufficient to support craftsmen, clerks, and soldiers, and western Europe was able to begin a counteroffensive against Islam. By the end of the century, Europeans had retaken a large part of Spain as well as the western Mediterranean islands and, with the collaboration of the Byzantines, they started the Crusades in their attempt to recover Palestine from the Moslems.

In the reconquered lands and on the borders, an extensive cultural interchange was going on with the brilliant, worldly, cosmopolitan Moslem civilization. Growing numbers of peaceful, inquisitive merchants and religious pilgrims visited Syria, Palestine, and Egypt. Shortly thereafter, so did the less peaceful Crusaders. Knowledge of and interest in the classical heritage began to spread throughout the West.

Revival of Learning

Belatedly, western Europeans began to appreciate the treasures in the captured Moslem libraries of Spain and Sicily. Belatedly because in Spain for three hundred years, from the time of the Moorish conquest, Christian monks, priests, physicians, and merchants had been familiar with Arab achievements and quietly passed information to their coreligionists in western Europe. These Christians, called Mozarabs, were Spanish and Portuguese natives who had adopted the Arab language and culture after the Moorish conquest but retained their own religion. Apart from the activities of the Mozarabs, not until the eleventh century was there any great effort to acquire Arab knowledge for Christian benefit.

The mental receptivity of the West was more important than its actual physical possession of Arab manuscripts and translations of classical Greek literature. After all, many Greek and Latin manuscripts already were in monastery libraries in the West and in Byzantium. These documents had been available for hundreds of years, but there was not much interest in them. Material conditions of life had to improve, and the struggle for existence had to be won before there could be time and energy for secular intellectual matters. Now many western Europeans had leisure to read and study, and contact with the Moslems had stimulated their intellectual curiosity.

In Spain, almost immediately after the reconquest of Toledo in 1085, a translation center was organized, employing bilingual Moors and Christian scholars and trilingual Jews. A Moor or Jew would read the Arabic document and mentally translate it, saying it aloud in Spanish, sentence by sentence, to a copyist who would write it down in Latin. In Sicily too there was a translation center where Greek, in addition to Arabic, texts were translated; educated Sicilians spoke Greek and Latin as well as Arabic. Actually, their knowledge of Greek enabled the Sicilian translators to compare Greek originals, when available, with Arab translations and so to correct errors made by earlier translators.

As a result, by about the start of the thirteenth century, more complete and correct versions of the writings of Aristotle replaced the earlier ones, and Aristotle rose swiftly to preeminence in western thought. In 1210 his works had been condemned and banned by the Church, but by about 1300, his position had begun to be reversed. By the end of the fourteenth century, Aristotle was the most authoritative of all philosophers, owing largely to the efforts of Albertus Magnus and Thomas Aquinas (1225–1274). With the flow of translations from Spain and Sicily into Christian schools and monastic libraries, for the first time in hundreds of years, Greek science and philosophy became generally available to the educated European. But Hellenic and Hellenistic science were not the only currents in the flood of information flowing into western Europe. The original works of the great Arabs al-Kindi, al-Razi, and Avicenna and the Jabir writings were also translated and so reached the clerics and physicians who had both the interest in the new material and the education needed to read and understand it. It took many years, however, for this information to trickle down to the largely illiterate artisans.

Cathedral schools continued to flourish and expand, and by the end of the thirteenth century, universities had evolved from them. There were universities at major cities such as Bologna, Paris, Padua, Naples, Siena, Salamanca, and Rome; at smaller towns such as Toulouse, Montpellier, and Orleans; and even at such hamlets as Oxford and Cambridge. These universities were, of course, oriented toward theology and to a great extent controlled by the Church. Some universities were organized and controlled by guilds of students, and others by guilds of teachers, but the original purpose of each had been to train teachers for the cathedral schools. (The doctorate was originally a license to teach.) Now new opportunities were opening up for secular careers in the state bureaucracy and in law and medicine. Education at Bologna became directed toward law and government service. In Paris, Montpelier, and Salerno, medicine became the specialty.

The Encyclopedists

From the beginning, lecturers at the new universities produced compendia of the translated knowledge as study materials for their students. At first, these compendia contained nothing new, no original work, just a rehash of what had long been known. Later on, as the lecturers added their own ideas, the combination of compendia and lecture notes gradually became textbooks. The first compendium of which we know was the *Liber de Proprietatibus Rerum* ("Book of the Properties of Things"). It was an encyclopedic work written around 1230 to 1244 by Bartholomew the Englishman, a friar who lectured at the University of Paris, as well as at other schools. The book summarized most of the Greek and Arabic works known at that time and had a wide influence, especially among clerical philosophers—including the Englishman Robert Grosseteste, who visited and perhaps studied in Spain, and his pupil, Roger Bacon.

Another encyclopedist was the Dominican Vincent of Beauvais, who died about 1265. He wrote the *Speculum Naturae* ("Mirror of Nature"), an enormous compilation of information that quoted over three hundred authors, many of whom are known only because he quoted them. The *Speculum* contained much chemical and alchemical information, all of it based on the opinions of Vincent's sources rather than on his own experiences. He used the Hellenistic–Arabic ideas of chemical principles, or essences, in classifying the many new substances.

Vincent accepted the theory of transmutation yet did not think it possible, in practice, to transmute metals. For transmutation, it would be necessary first to change the base metals into the prime matter and then to reconstitute them into gold. Nature could do that, but humans were not powerful enough. Therefore, they could only color metals, not transmute them. Vincent was not only skeptical about transmutation but probably cynical as well.

The prolific writer Albert von Bollstädt (1193–1280) was another influential clergyman–encyclopedist. He was the Bishop of Regensburg, called by his contemporaries Albertus Magnus ("Albert the Great"). He was quite interested in esoteric alchemy, as was his younger contemporary, Thomas Aquinas, but like Vincent, he was a compiler and not an experimenter. Albert was a great popularizer, writing clearly and directly with the purpose of disseminating knowledge, and his works were very influential in spreading doubt that alchemists could actually perform transmutations. A collection of his works, published in Paris from 1890 to 1899, ran to thirty-eight volumes.

The Decline of Alchemy

After centuries of unrelieved failure to produce gold in any quantity, there was considerable skepticism about exoteric alchemy and alchemists. The development of solutions of nitric acid, hydrochloric acid, and aqua regia also raised doubts about alchemy. Lead could now be analyzed routinely with acids, and it was realized that gold was already present in lead even before any attempt at transmutation. This realization in itself was not a complete surprise; according to accepted theory, all metals changed spontaneously, but slowly, into gold. Therefore, all metals were expected to contain some gold. The real surprise was the observation that, after transmutation, the weight of gold in lead was no greater than it had been before transmutation. After mineral acids were applied to the analysis of metals, fewer and fewer natural philosophers believed in transmutation by human agency. Some alchemists, however, still kept their faith. After spending everything they had in fruitless attempts to produce gold, they wandered from place to place, trying to find

An alchemist making an elixir. The lion swallowing the snake symbolizes the melting of a metal. (Reproduced with permission from the Bettmann Archive.)

a wealthy patron to finance more experiments. If they did find backing, when the inevitable failure occurred, their only safety was in flight. Times were hard for gold-seeking alchemists.

The Church in particular was strongly opposed to alchemy. For one thing, the doctrine of the elixir of life was sacrilegious and contrary to Christian dogma. Eternal life was not to be confused with living forever. It came through the sacraments of the Church and not through any magical potions. Moreover, from its very inception, alchemy had been connected with paganism and magic rites. About the year 1400, the Christian world started a three-hundred-year war against witches, wizards, the remnants of the pagans, and anyone suspected of practicing magic (*1*). Scores of thousands of pathetic, senile, mentally defective, or emotionally disturbed people who mumbled to themselves or hallucinated were burned at the stake. The butchery extended to helpless animals, including thousands of cats. Pet cats, especially black cats, were considered to be witches' familiars, and were tortured, burned, and killed on sight. Even today, there is a superstition that black cats are a sign of bad luck.

In such an intellectual climate, the alchemist had much to fear. Around 1300, Dante, in *The Divine Comedy*, pictured alchemists being tortured in hell. In 1317 Pope John XXII issued a decree against alchemy. In 1323 all clergy who practiced alchemy were excommunicated. In 1380 the King of France ordered persecution of alchemists and forbade anyone to own alchemical equipment. In 1404 in England, Henry IV banned alchemy, and in 1418 even worldly, sophisticated Venice forbade its practice. (Actually the practical Venetians were probably more concerned with counterfeiting than witchcraft.) The troubles of witches and alchemists persisted for centuries. Witch hunts in Europe did not begin to die down until the mid-seventeenth century; the last reported European execution for witchcraft was in Poland around 1790. Still, even in the worst of times, there were always plenty of alchemists trying to transmute lead into gold. No law or persecution could prevent people from trying to get rich.

Roger Bacon

The Franciscan Roger Bacon (c. 1214–1294) differed from his predecessors Albertus Magnus and Vincent of Beauvais in that he was not only a writer and a popularizer of alchemy but also an experimenter. In the year 1267 he wrote that he had spent a large sum, the equivalent of many thousands of today's dollars, on alchemical books as well as on instruments and equipment. Moreover, in his writings, he specifically stressed the importance of experimentation. In fact, he refused to accept theory without experimental corroboration. To Bacon, however, experiments were not intended to discover new phenomena or new principles or to develop new theories; they were designed to demonstrate the validity of theory deduced from Scripture and accepted general principles.

Perhaps influenced by Franciscan physicians (*see* Chapter VII) or the Chinese, Bacon considered the most important uses of alchemy to be healing the sick and studying how to prolong life indefinitely, forever if possible (2). He believed that people die because of corruption. Distilled medicines should remove corruption, and alchemy was, therefore, a valuable source of medication. This idea flatly contradicted the accepted beliefs of medieval physicians. Loud echoes of this thinking appeared in the work of another important Franciscan, John of Rupescissa, some fifty years after Bacon's death.

Bacon got into trouble with the ecclesiastical authorities and was either imprisoned or put under house arrest for fourteen years for reasons that are unclear to us. Probably his prickly personality had something to do with it. Like Leonardo da Vinci, he predicted many inventions and phenomena that later came to pass, although some of Bacon's predictions were simply figments of his imagination. However, he was the first European to write about gunpowder, and that was certainly not a flight of fancy.

Gunpowder

Gunpowder had been used by the Chinese for centuries, not in guns but in incendiary grenades. In 1240, Bacon first mentioned gunpowder, and by 1324, or even earlier, the inventive Europeans had learned to confine the powder in a closed space, so that the force of the explosion would propel a projectile. They had invented guns and cannons.

In a 1327 book by Walter de Milemete titled *De Nobilitatibus, Sapientibus, et Prudentiis Regum* ("Of the Nobility, Wisdom, and Prudence of Kings"), there is a picture of a cannon, in the shape of a vase or an inverted bell, firing an arrow against a castle. By 1480, gunpowder was being used in the Alps in road construc-

A bomb or grenade bursting near a Japanese bowman. (Reproduced with permission from reference 3. Copyright 1956, Oxford University Press.)

tion. And by then, it was, of course, extensively used in warfare, bringing to an end the military supremacy of the knight in armor.

From 1350 on, no European ruler could afford to be without supplies of gunpowder, which meant also supplies of sulfur and saltpeter. Therefore, no matter how hostile the Church and King might be to alchemy in general, those alchemists who could make gunpowder were welcomed and protected.

Late Medieval Alchemical Writers

Around 1311 a collection of writings appeared by an alchemist, most likely a Spaniard, who for safety's sake published under the name "Geber," a latinized version of Jabir (*see* Chapter IV). His books were among the most influential of western alchemical writings. The best known was the *Summa Perfectionis Magisterii* ("The Perfect Reagent"), which is essentially a scholastic treatment of alchemy (4). It deals with both theory and practice and contains chapters on testing, mostly of gold and silver. In it, Geber modified contemporaneous beliefs by suggesting that metals consist of arsenic as well as sulfur and mercury, but this idea was not generally accepted. Geber's methods included reaction with "corrosive sublimate," our modern mercuric chloride, which can act as a chlorinating agent. He also mentioned and discussed saltpeter, nitric acid, vitriol, and other materials of increasing importance in the newly emerging technology. He believed in transmutation by the elixir and considered that in a successful transmutation the alchemist's task was to purify the transmuted metal by removing excess sulfur and earth with heat, acids, and distillation. His works would obviously be expected to lay great stress on laboratory methods of purification. In fact, his books were really laboratory manuals.

Geber was systematic, clear, and concise, with an apparent desire to explain experimental methods and observations. His great influence, like that of Albertus Magnus, was in his level-headed attitude, although unlike Albertus he was an experimenter and not just a compiler. For example, he wrote about the purification of soda (sodium carbonate):

> Soda is purified like common salt. First it is ground and entirely dissolved in warm water, afterwards filtered and solidified [that is, cooled until it precipitates out of solution] and calcined with gentle fire [heated to drive off the water of crystallization] [5].

These descriptions are clear and practical. Contrast them with some of the previous chemical writings, such as in the Stockholm papyrus (*see* Chapter III).

Two other famous writers of the period were Arnold of Villanova (c. 1235–1311), a physician who used chemical remedies and had a great reputation for cures, and Ramon Lull (1232–1315), a medical friar who was also a missionary. Actually, neither man was an alchemist; the alchemical works attributed to them were probably published centuries later by alchemists hiding under their names. Arnold did use mercury and arsenic compounds and alcohol as medicines, and in that sense, he was a predecessor of Paracelsus, the great chemotherapist.

Europe from 1330 to 1550

After about 1330, for some two hundred years, no real changes took place in chemical practice. Artisans continued to improve existing processes on a trial-and-error basis and to develop new processes and chemicals, while the alchemists kept a low profile in the face of Church hostility. Except for the introduction of gunpowder, until about 1500 there were few significant changes other than a deepening interest in medical alchemy. This long hiatus probably stemmed from depressed economic and disorganized social conditions in the wake of wars, devastating plagues, famines, and revolutions. Two centuries of suffering discouraged natural philosophy and science.

The Hundred Years' War between England and France eventually involved almost all of western Europe. There were other wars too. In England, the Wars of the Roses (the red rose being the badge of the House of Lancaster and the white rose that of the House of York) were savage civil wars in which the English nobility virtually exterminated itself, leaving the way clear for the Tudors and the New Men of the Renaissance. Trade wars between Barcelona, Genoa, and Venice spilled over into the entire Mediterranean area and the Balkans, accompanied by brutal fighting, piracy, pillage, the sack of towns, and indiscriminate massacre.

Among the catastrophic events were bitter class struggles. In the industrial towns, the individual craftsman had now become a pieceworker rather than an independent tradesman, losing status and economic strength and often becoming impoverished. In the fields, war and a succession of bad harvests had reduced the peasants to destitution and desperation. At the same time, religious fervor had largely decreased. The early Church had enjoined poverty, love, humility, and chastity on monks and priests, but now the Church itself was a great temporal power, a great landowner, and a consumer of luxury goods. The princes of the Church were wealthy magnates. By late medieval times, the gap between rich and poor and the difference between the theory and practice of the Church often resulted in exasperated class struggle with all the manifestations of hatred, often directed against the Church itself.

Flanders, the great center of medieval industry, was in the forefront of the conflicts. In 1274 the textile workers of Ghent struck and abandoned the town. At about the same time, the ironworkers in Liege and the copper beaters in Dinant rebelled. In 1280 and again in 1298 the entire industrial population of Flanders rose up in a general strike that became a devastating civil war. The peasants rebelled again between 1323 and 1328, supported by the industrial workers of Brugge and Ypres. In France in 1356 there was another peasants' revolt, or jacquerie.

Usually there were about twenty years between outbreaks in any particular locality, enough time for a new generation to grow up and make a fresh attempt. The rhythm was the same as that of the nineteenth century outbreaks of revolution in Europe. In 1378 and 1382 came fresh revolts in Flanders and France, and now the unrest radiated outward in all directions. In the South, the city workers of Florence, the Ciompi, seized power for a time. In the North at Lübeck on the Baltic and in the East at Prague, bloodbaths occurred. Neither side in these struggles showed any mercy.

In England in 1381 the peasants broke out in Wat Tyler's Revolt, sparked by the egalitarian preachings of a radical priest, John Ball. Tyler actually gained control of London before he was murdered during a meeting with the young King Richard II. In 1415 in Bohemia, after Jan Hus was burned at the stake, the peasants revolted against both Church and state. By the mid-fifteenth century, the revolts had spread to Scandinavia and Finland.

Many of these uprisings were directed at the wealth of the medieval Church, but even after the Protestant Reformation of the sixteenth century, when the wealth of the Church was confiscated in Protestant regions, class warfare continued. Peasant uprisings became even more revolutionary. In fact, Martin Luther lost much popular support when he sided with the princes during the great Peasants' War in sixteenth century Germany.

More catastrophic than war and revolution was the Black Death, the bubonic plague. Starting about 1347 in Naples, Genoa, and Constantinople, the Black Death ravaged Europe. It died down but broke out again frequently until about 1500 and then at longer intervals thereafter. (Even today, bubonic plague is endemic in some regions, notably India and California, where it is

The earliest known representation of a glass-enclosed analytical balance from a 1477 book by Thomas Norton, Ordinall of Alchemy. *Balances like this were standard equipment until the 1950s. (Reproduced with permission from the Bettmann Archive.)*

spread by the ground squirrels.) In 1349 as much as a third of England's population died. In the towns of Europe, between 1347 and 1352, perhaps two-thirds of the population died. Hamburg lost two-thirds of its population, Bremen about seventy percent, and Genoa slightly less than two-thirds. The mortality rate was not so high in the rural areas, but even there whole regions were almost depopulated.

Finally, there was famine. Famine not only killed countless people but contributed to the ravages of the plague by lowering resistance to infection. Between 1314 and 1317, three consecutive crop failures caused death and widespread suffering over the whole of Europe. In 1316 at Ypres in Flanders one-tenth of the population starved to death in a six-month period. Famine struck again and again, between 1346 and 1350, 1361 and 1362, and 1374 and 1375.

The grim picture painted here seems at first to be contradicted by the tradition of the contemporaneous Italian Renaissance, an era of love of life; of awakening interest in learning; of improvements in education; and of tremendous interest and advances in architecture, literature, and the fine arts. In reality there is no contradiction. The same conditions prevailed in Italy as in the rest of Europe, although in Italy they were mitigated by relatively favorable circumstances. The wars were, for the most part, in northern Europe. Italian trade continued with less disturbance than that of the bellicose countries. The famines too were less severe. Basically, however, the Italian Renaissance does not give a true impression of economic and social conditions because it was an aristocratic movement. It was financed and supported by wealthy churchmen and magnates. The Medici, the Borgias, the Dukes of Ferrara and Urbino, and the Visconti, Farnese, and Sforza families were the great patrons of the Renaissance artists and humanists in Italy.

Only in the sixteenth century did the renaissance in literature and the arts spread to the North, where the patrons of the arts were merchants as well as princes. The portraits of opulent Dutch and Flemish burghers give a true picture of middle-class prosperity in the sixteenth and seventeenth centuries. The absence of such paintings in the fourteenth and fifteenth centuries tells us something else.

The combined effect of pestilence, war, and famine had produced an intellectual climate in which slow, patient scientific investigation was not particularly appealing, to say the least. Progress in industrial technology did continue, especially as applied to armaments and improvements in mining and assaying. But above all, in plague-stricken Europe, there was a heightened interest in medicine. The next advance toward chemistry was to come in medical alchemy.

References

1. Trevor-Roper, H. R. *The European Witch Craze of the Sixteenth and Seventeenth Centuries*; Harper and Row: New York, 1956.
2. Stillman, J. M. *The Story of Alchemy and Early Chemistry*; Dover: New York, 1960; p 263. Reprint of 1924 edition, Appleton: New York.
3. Singer, C.; Holmyard, E. J.; Hall, A. R.; Williams, T. I. *History of Technology*, Vol. II; Oxford University Press: London and New York, 1956. After Goodrich, L.; Chia-Sheng, F. *Isis* **1946**, *36*, 118.
4. Multhauf, R. P. *The Origins of Chemistry*; Franklin Watts: New York, 1967; p 172.
5. Stillman, J. M., op. cit., p 281.

Timeline—Alchemists and Natural Philosophers

	1085–1150	1150–1200	1200–1300	1300–1327	1328–1380	1380–1450	1450–1500
Technological and Scientific Advances	Bologna University founded, 1119	Oxford University founded, 1167	Bacon mentions gunpowder in writings, 1240 Vincent de Beauvais writes *Speculum Naturae,* 1250 Albert von Bollstädt (Albertus Magnus) lived, 1193–1280	Geber's works appear, 1311 Cannon used in sieges, 1327	Pope John XXII bans alchemy in 1317 and repeats ban in 1323 Alchemy banned in Venice, 1380	Alchemy banned in England, 1404 Alchemy banned in France, 1418	
Historical Landmarks	Toledo recaptured from Moslems, 1085 Crusades, 1096–1291			Recurrent famines in Europe, 1314–1375	Hundred Years' War between England and France, 1337–1453 Onset of Black Death, 1347	Jan Hus burned at stake, 1415	Wars of the Roses in England, 1455–1485

VII
Medieval and Renaissance Medicine and Medical Alchemy

The contributions of physicians and medical alchemists to the development of chemistry have been seminal. It is axiomatic that physicians played a major role in the origin and growth of early chemistry. For hundreds of years, until the end of the eighteenth century at least, most of the great discoveries in chemistry were the work of physicians. They had excellent scientific training for the era, the money to set up laboratories, and the time to spend on experimentation. In fact, when chemistry was first taught at the university level, it was as a service course in medical schools. Less well known than the contributions of the physicians, but perhaps equally important, are those of the medical alchemists. They discovered many important chemicals and chemical reactions. By the mid-sixteenth century, their work had helped change the mindset of those dealing with chemicals and made them receptive to the ideas of atomism and chemical composition.

Medieval Medical Care

In medieval Europe, most sick people were treated either by prayer or by folk medicine and sympathetic magic. Although a relative few—kings, nobles, high churchmen, bankers, and wealthy merchants—could also call on the services of trained physicians, the first resort was usually prayer. The priest would bless and pray over the patient, and when available, holy relics, such as saints' bones, would be touched to the site of the pain. For continuing illness, the patient might visit a shrine, such as that of Saint James of Compostela, just as people do today at Lourdes or Fatima.

If prayer or pilgrimage didn't help, those who could afford it would call in a physician. The few professionally educated physicians had been trained in the system of Hippocrates (c. 460–370 B.C.), as modified by Galen (c. 129–200 A.D.). They believed that disease was caused by an imbalance of the four body humors: blood, black bile, yellow bile, and phlegm. Their theory was largely harmless, except that it kept physicians from considering external agents as causes of disease.

The Hippocratic treatments were intended to restore the balance of body humors by physical means and by drugs. For example, a patient with chills would be placed next to a roaring fire. Mostly the physical treatments had no great effect, except when it came to bleeding, a practice that continued as late as the mid-nineteenth century. If a condition was attributed to too much blood, the patient was bled to remove the excess. Unfortunately, the patient was often bled to the point of collapse, "bled white." One notable case is that of George Washington, who in 1799 was fatally weakened by being bled for a common cold.

Hippocrates' medications were obtained from plants and minerals and were usually prescribed in small quantities that had little effect. Galen modified Hippocratic medications in a

A thirteenth century English illustration of a physician weighing out an ingredient while he supervises his assistants. One pounds drugs in a mortar and the other stirs a pot. (Reproduced with permission from the Bettmann Archive.)

manner that made them dangerous and often deadly. He added materials of animal origin that were in many cases foul, disgusting, and even potent sources of infection. As late as 1618, the *London Pharmacopeia* listed many Galenic ingredients, among them: human and animal urine, excrement, and sperm; blood, bile, and intestinal contents; worms; lice; ground bones; bird droppings; snake meat and venom; and scorpions. Moreover, Galenic medications were composed of scores of ingredients, so it was impossible to tell if any particular component was helpful, ineffective, or even positively harmful.

Those patients who could not afford or did not trust physicians would go to the local seer, a "wise" man or woman who practiced sympathetic magic.

Sympathetic magic dates back to the Stone Age when the idea arose that objects were magically connected with their names or their images. Uttering the name or creating an image gave supernatural control over the person or object. If a caveman painted a picture of a deer with arrows in its body, out in the woods the hunters would be able to kill a deer with arrows. If a wax image of a man were stuck with pins, the man would die. Sympathetic magic was widely practiced in medieval Europe. The term *cancer* comes from the Latin word for crab, so if a man had a tumor, a crab would be tied to the site of the pain or the tumor and killed. As late as the seventeenth century, in enlightened England, even among the educated, the belief persisted that if a man had been wounded by a sword, putting an ointment or a dressing on the sword would heal the wound. Such magical remedies, while ineffective, at least did no harm, in contrast to the lethal drugs of the physicians. Essentially, the patient of a folk practitioner died of the disease, and the patient of a physician died of the cure.

Medical Problems

The medical problems faced by the medieval world were staggering. Crowded, unsanitary, vermin-ridden conditions and contaminated food and water kept life expectancy at a low level. Preventive medicine did not exist. Sanitation was primitive. Kings and queens perished of typhus, transmitted by lice, and of bubonic plague, transmitted by fleas; even kings had fleas and lice. Proper nutrition was rare. As a rule of thumb, women lost one tooth with each child after the first, because of calcium deficiency. As late as 1600, Shakespeare's Juliet was ready for marriage at the age of thirteen. This may have been poetic

Death and the newly married lady, from a 1538 woodcut by Hans Holbein, the youngest. Note the man's pock-marked face, characteristic of syphilis. (Reproduced with permission from the Bettmann Archive.)

license, but marriage at the onset of puberty was common. Not too many women lived past thirty-five, many dying, especially in childbirth, worn-out, toothless old hags in their late twenties or early thirties.

In the ancient and medieval world, and until the nineteenth century, most people died of hunger or of bacterial or viral infections. Even during warfare, more soldiers died of dysentery than of wounds; World War I was the first major war in which more soldiers were killed in battle than died of disease. Hunger lowered resistance to disease, and malnutrition, even among those not actually underfed, left many people too weak to survive infection. Disease of internal organs, such as appendicitis, mastoiditis, stomach ulcers, and prostatitis, was almost invariably fatal. Without anesthesia, the patient frequently died of surgical shock during an internal operation. There was not the slightest idea of asepsis or disinfection, so if the patient did survive the operation, post-operative infection was usually fatal. People seldom lived long enough to die from cancer and heart or circulatory disease. Gout was considered to be a disease of the wealthy, because almost everyone else was on a low-protein diet. Simple and compound fractures could be set, and there was considerable experience with these and with combat wounds. Eye disease, especially flyborne, was widespread throughout the Mediterranean area, although not so much in the northern climates. Skin diseases were especially common because of the lack of soap for removing ground-in dirt.

The great killer diseases were the bacterial and viral infections. Infected cuts and wounds were often fatal. Even a slight scratch could be dangerous because there was no way to sterilize a cut except with cautery, in which case the resulting burn often became infected. Farmers used human and animal manure for fertilizer, so tetanus, typhoid, cholera, and bacillary dysentery were ever-present dangers. Because of the almost total lack of sanitation, gastrointestinal diseases were everywhere, even in palaces. Wastes were dumped into water supplies. In most cases, people ate from common pots, dipping in their hands or spoons. (In seventeenth century France, Louis XIV ate stew with his fingers.) There were great outbreaks of cholera and typhoid. Diseases with long incubation periods, such as plague and typhus, caused havoc. Diseases with short incubation periods could devastate a locality, but travel was so slow that an outbreak was not likely to spread to another area. The sick traveler would either recover or die before he got to the next city. (Today, epidemiologists have to worry about the spread of diseases from one continent to another at supersonic speeds.) Respiratory diseases had no cure, other than the natural resistance of the patient. Here again, the difficulties of travel usually restricted outbreaks to small areas. However, when an outbreak of respiratory disease did strike a town, it was devastating because of the crowded, unsanitary conditions.

Until the development of antitoxins and vaccines in the nineteenth century and antibiotics in the twentieth century, the only effective treatment for infectious diseases was dosing the patient with inorganic substances or with alcohol. The alcohol was effective as a cleansing agent externally and as a mild bactericide. The inorganic remedies were solutions and ointments of powdered minerals, usually containing heavy-metal compounds. Unfortunately, the orthodox, or Galenic, physicians of the medieval and Renaissance eras used neither heavy metals nor alcohol to any great extent.

The Franciscans' Medicine

The first faint beginnings of improvement in treatment of the sick came about the year 1200, when Franciscan friars began to practice folk medicine. For priests, monks, and nuns, helping the sick was an important part of their religious duties. Franciscan friars, who were especially dedicated to living in the secular world and helping others, both gave and received charity and devoted themselves to ministering to the sick. Living as they did among the poor and hungry, who were mostly peasants, they tended to use folk remedies and sympathetic medicines. But, in a move of immense significance, they added alcohol to their list of remedies.

Priests and monks had long experience with sacramental wines and distillation and became aware of alcohol as soon as it was discovered. Struck by its startling properties, and using the logic of sympathetic magic, the Franciscans reasoned that the distilled alcohol had been purified by the fire and freed from the dirty brown fermentation mash just as the soul was plucked by the true faith from the dirt and dross of the temporal world. Therefore, purified alcohol would in turn purify the patient and cure the disease.

Treating patients with alcohol actually did help, and alcohol soon was widely recommended as a panacea. (Arnold of Villanova, for one, built his reputation on alcoholic medicines.) Washing wounds with alcohol cleansed them and killed some microorganisms. More important, administering alcohol to the patients made them drunk—relaxed, comfortable, perhaps even happy. The patients were spared the foul, dangerous Galenic potions, and the body had a chance to heal itself. The Franciscans soon acquired a reputation as healers.

Around 1100 A.D., the idea of the elixir of life reached the

Duffy's Pure Malt Whiskey for Medicinal Use. Sold by all Druggists, and bears the highest endorsement of the leading chemists and medical authorities of the Country.

Duffy Malt Whiskey Co.
ROCHESTER, N. Y.

For Sale by
SMITH, KLINE & FRENCH CO.
PHILADELPHIA, PA.

A nineteenth century version of the elixir of life. (Reproduced with permission from the Bettmann Archive.)

West. In the early 1200s, the Franciscan medical practitioners began to establish a connection between alchemy and medicine, possibly deriving the connection from Chinese alchemists, who wanted to use gold only for medicinal purposes and sought the elixir. By 1267, the Franciscan monk Roger Bacon was studying medical alchemy as a way of saving bodies in addition to souls. Medical alchemy, unlike esoteric alchemy and transmutation, was becoming respectable. Bacon declared that the true purpose of alchemy was not to make gold for financial gain, but to make distilled medicines that would overcome the corruption that caused illness.

Chemotherapy

About fifty years later, another Franciscan, the great and almost forgotten John of Rupescissa, a medical alchemist and a medical missionary, acted on Bacon's ideas (*1*). John was an idealistic and visionary Catalan who was imprisoned several times for inflammatory preaching against the Church hierarchy. He believed that the cause of disease was a corruption of some kind, and that the great purpose of alchemy was to make medicines to overcome this corruption and save lives. He sought a quintessence from which to make the elixir. In his book, *De Consideratione Quintae Essentie* ("Considerations of the Fifth Essence"), he stated that alcohol was the quintessence of which the heavens are made, but in most of his other writings he seemed to feel that alcohol merely contains the quintessence. In fact, to John, all things contained the quintessence. It could be extracted from organic and even mineral material by distillation and sublimation, that is, by alchemical procedures.

He, and later on his followers, took all sorts of mineral and organic materials such as bricks, excrement, and urine and treated them with acids and alkalies, roasted them, digested them, mixed them with whatever they could get their hands on, and finally distilled or sublimed the mess. And they often ended up with relatively pure chemicals, some of which had been unknown previously. Over the next two hundred years, the medical alchemists, eventually known as iatrochemists, not only found scores of new substances but also prepared previously known chemicals in pure form.

The substances obtained were immediately tried out as medications and many turned out to be effective. Ammonia from distilled urine became a standard medication. So did antimony compounds. Antimony sulfide had been known to the Chinese, Egyptians, and Arabs for thousands of years as a treatment for stomach pains and skin irritations, but in the West it was apparently first prescribed by John of Rupescissa. This coincidence suggests a possible link with the Chinese, because antimony sulfide may well have been the potable gold mentioned in Chinese alchemical works.

Although we have no direct evidence, John was probably well aware of the Chinese idea that drinking gold would prolong life. Europeans, especially missionaries and sailors and the traveling traders of Venice, Genoa, and Barcelona, knew more about China than is generally realized. In 1245 the Pope sent the Franciscan

John of Piano Carpini to the Mongol court as an ambassador (2). Other clerical ambassadors, usually Franciscans, followed, and there was a constant Christian presence at the Chinese Imperial Court until the fall of the Mongol (Yuan) dynasty in 1367. Moreover, throughout the thirteenth century, merchants from the great Mediterranean seaports traded extensively with Persia, which was ruled by the Mongol Il–Khans after 1218. There they undoubtedly met traders from China, the eastern part of the Mongol empire. European traders, such as Marco Polo and his family, also traded in China itself. In turn, the Mongols sent traders and envoys west to Rome, Paris, and Constantinople (3). All in all, it would be surprising if in 1300 John of Rupescissa, a Franciscan from the seaport of Barcelona, had not learned something about China from merchants and sailors, from the reports his fellow Franciscans were sending home from the Chinese court, and from the writings of Marco Polo.

In any case, John introduced his own version of ingesting potable gold to cure all bodily ills. He combined the idea of the healing powers of gold with his belief that alcohol was either the quintessence or a quintessence when he produced what he called "potable gold." He heated gold or silver, dipped it into alcohol, and served the alcohol to his patients. Gold does not react with alcohol, so all that he really administered was warmed alcohol with perhaps some dirt or grease dissolved in it. Nevertheless, John's potable gold had a great vogue, and even today something called potable gold is being sold in health food stores.

Chemical Medications

The advent of alchemically prepared medications presented the alchemist with both a great opportunity and a horrendous problem. Medical alchemists were relatively free from being under suspicion and from the persecution that other alchemists endured. Medical alchemy was therefore an acceptable trade, a safe way of earning a living. On the other hand, treating people with alchemical medicines was a risky business. Unless the physician knew his materials, he was playing Russian roulette with his patients' lives. Some substances fed to patients cured them, quickly and dramatically. Others, superficially very similar to the cures, killed them. If a physician fed a patient a white powder and it turned out to be corrosive sublimate, the patient would die slowly and painfully. Arsenic compounds were poisons, but antimony and bismuth compounds could be cures[1]. Because arsenic, bismuth, and antimony compounds not only have similar properties but also are often found mixed together in natural deposits, bismuth or antimony medications were often contaminated with deadly amounts of arsenic. It was literally a matter of life or death for the iatrochemist to be able to separate mixtures into their components and to purify and identify all materials present.

Analytical separations and purifications had to be worked out for all medical chemicals, and lists of identifying physical and chemical properties had to be prepared. The medical alchemists had to become, and by the sixteenth century did become, practical analytical chemists. They acquired an empirical knowledge of chemicals and chemical processes that would have staggered the natural philosophers and physicians of Aristotle's day. They learned the properties of scores of chemicals and the courses of hundreds of reactions. In separating reaction mixtures into their components, they found that almost invariably one of the starting materials was still present. The Greeks had been wrong in assuming that materials could combine in any proportion. They found also that the properties of pure materials usually stayed constant; therefore, the elements were not really changing back and forth into one another. In fact, they learned to rely explicitly on chemical and physical properties to identify the material. Finally, more and more, the iatrochemists worked with purified chemicals in their reactions, which may not sound like much of an advance, but it was. Working with relatively pure chemicals nurtured the feeling that chemicals were made of other chemicals, not philosophical elements or essences or principles. By about 1500 A.D., many iatrochemists and natural philosophers familiar with their work had developed doubts about Aristotle's chemistry. They were becoming dissatisfied with the Four Elements theory but had nothing to put in its place.

By that time, the climate was also ripe for a change in medical viewpoint. Even orthodox physicians were moving away from the complicated Galenic treatments toward alcoholic extracts and simpler medications prepared by sublimation and distilla-

[1]The difference depends on solubility. Arsenic salts are all quite soluble in gastrointestinal fluid, and thus the ingested arsenic reaches a high concentration, so high that it passes through into the bloodstream and poisons the victim. Bismuth compounds are much less soluble, and so the concentration of ingested bismuth is quite low, high enough to kill intestinal microorganisms but not high enough to enter the bloodstream in any appreciable concentration. Bismuth is still used today in prescriptions and patent medicines such as Pepto-Bismol for upset stomachs and gastroenteritis. Antimony has a solubility midway between that of bismuth and that of arsenic and it ranks midway between them in terms of medical effectiveness.

tion. Not only had the alchemical physicians achieved many cures, but orthodox physicians had been helpless against the great killers, plague, smallpox, and syphilis. Galen himself had begun to suffer a loss of reputation when, in 1542, Vesalius published his great work *De Humani Corporis Fabrica* ("Concerning the Architecture of the Human Body"). Its exquisite illustrations showed that Galen had made anatomical errors. At that point, even the Galenic physicians were ready for change. Perhaps medical alchemy could have merged quietly with orthodox medicine over the next few generations, with Galenic potions gradually being abandoned and the medical alchemists becoming accepted as physicians. Still, physicians were more than just another guild of specialists. They were a privileged caste, with great wealth and social status that they could lose if alchemists were permitted to promote their own theories and cures. In any event, the personality of that extraordinary medical alchemist, Paracelsus, was more than enough to start a revolution.

Paracelsus

Almost all details of the life of Paracelsus are shrouded in doubt and the controversy that he constantly provoked (*4, 5*). He was Swiss, born about 1493, and died in 1541. Paracelsus was a nickname, probably given to him by others and adopted by him because he claimed to be superior to and beyond, or "para," to Celsus, the Hellenistic poet and supposed physician. His name was Philippus Aureolus Theophrastus von Hohenheim, although sometimes the name Bombastus is included. Bombastus has nothing to do with the adjective bombastic, although his writing style certainly warranted the description. His paternal grandfather was of the Bonbast or Bombast family of southern Germany. Von Hohenheim did not signify nobility, Hohenheim being the town from which his father came.

His father had a medical license, although not necessarily a medical degree. Paracelsus had a classical education and was proficient in Latin, although he wrote and lectured in German, which was itself an innovation. He is supposed to have studied at various schools and also to have wandered around Europe, picking up information from the people with whom he worked. He spent some time working in the mines in southern Germany, where in 1516 he was apprentice to an assayer, analyzing ores for metal content. There he learned assayer's alchemy, the equivalent of qualitative and quantitative inorganic analysis, becoming a very competent laboratory technician. He studied medicine at

Ferrara, in Italy, and claimed to have obtained an M.D. degree there. This claim has not been substantiated, although Paracelsus did have a thorough knowledge of medicine. Probably he studied there but did not graduate. In 1522 he worked as an army surgeon during one of Europe's incessant wars. The experience was instrumental in his dislike of physicians, who considered surgeons to be unlettered artisans and treated them disdainfully. In spite of his later claim to the medical degree, in 1526 he enrolled in the guild of grain merchants at Strassburg rather than in the physicians' guild. He called himself a doctor of medicine; in his books, a doctor of theology; and in his will, a doctor of liberal arts. In any event, he was well grounded in alchemy and in practical medicine as well as in the classics of medical theory that he rejected. He wrote, or rather dictated, to Operinus, his factotum, an enormous number of books, of which twenty-two are of chemical interest.

In 1527 he was appointed a medical health officer at Basel, a position that did not require a medical degree. He was to lecture on medicine, although not on alchemy, and to supervise apothecaries. One widely circulated but improbable story states that he began his lecture series by burning the works of Avicenna (or perhaps of Galen), which would have cost him a small fortune. In his lectures, he abused conventional physicians in scatological terms. He not only objected to their remedies and the very basis of their medical treatments, but attacked even their life styles. He is supposed to have passed a pan full of excrement around the audience and then railed furiously at those who were offended and left. Whether or not this incident is true, Paracelsus was continually in conflict with the medical establishment. He wanted the physician to be a surgeon and the surgeon a physician. He wrote, in his typical apoplectic style, "If you are not a physician, what can you do other than mechanical cutting in tailor's fashion?" and "When the physician is not a surgeon, he is an idol that is nothing but a painted monkey."

During his short stay in Basel he earned a great reputation by curing the famous printer Frobenius, who was a friend of the Reformation scholar Erasmus. The illness, the treatment, and the results are, as usual, obscured by uncertainty and speculation. Frobenius is said to have suffered from gangrene for two years as a result of a fall that injured his leg. Paracelsus treated him and restored his health to such a degree that about six months later Frobenius undertook a long journey. Paracelsus warned him against going on the trip and told him that he would have difficulty staying on his horse. Sure enough, Frobenius fell off his horse, struck his head, and died. Paracelsus's detractors claimed that he had not been cured. His supporters argued that Frobenius had indeed been cured and had died as a result of an avoidable accident. In any event, gangrene caused by an infection could not have lasted two years, and Paracelsus could not have cured it. One possible explanation is that Frobenius suffered from a circulatory problem, such as that caused by diabetes, and was developing an ulcerated condition that was aggravated by his original fall. What Paracelsus did for the underlying problem we simply do not know, but he probably sedated the pain with opium, a favorite drug. Opium, or any other sedative, would have made prolonged horseback riding dangerous, because a sedated man could not be expected to maintain his balance on a horse for long.

Paracelsus's career in Basel soon came to an abrupt and characteristic end. For a fee of one hundred gulden, he cured a magistrate of gout, again probably by deadening the pain with opium. When the cured magistrate refused to pay, Paracelsus sued him in the magistrate's own court. As might be expected, Paracelsus lost the case and roundly denounced the magistrate, the court, and the town authorities. His friends managed to get him out of town, but he spent the last years of his life wandering all over northern Germany, not settling down anywhere, even though he was offered desirable positions as a physician.

Paracelsus's Ideas

Through all of the charges, claims, and invective hurled by his opponents and supporters, it seems clear that Paracelsus had both a brilliant intellect and an abrasive personality. Perhaps he was even mentally ill. He must have been an excellent medical practitioner. Even his opponents admitted that he achieved many cures. In the sixteenth century, he wrote a book on the industrial diseases of miners, among the first books on environmental medicine, in which he discussed the harmful effects of mercury and arsenic vapors. He stressed the physician's obligation to the patient. He used soporifics, sedatives, and narcotics, including opium, to alleviate his patients' pain and to make them comfortable. He treated his patients with what apparently was diethyl ether ("sweetened vitriol" made by distilling mixtures of alcohol and sulfuric acid). He also demonstrated the effects of ether on laboratory animals three hundred years before Crawford and Morton in nineteenth century America pioneered the use of ether as an anesthetic. He emphasized purification of his medications, usually by distillation. He was careful in controlling dosages, remarking correctly that even poisons could be beneficial if given in small doses. He believed that the body's own resistance can overcome a disease or heal a wound, and stated that the most important thing a surgeon can do is to keep the wound clean while nature heals it. He believed that each disease was different from every other disease and could be cured only by a "specific" medicine; in fact, he was the first to use the term specific to refer to such a medicine. He and his followers looked for specifics, especially for those that contained only one ingredient. Therefore, it was relatively easy for them to see which were effective, ineffective, or harmful. As a result, their cure rate was much higher than that of the orthodox Galenic physician.

Nevertheless, Paracelsus's abrasive personality and confrontational style alienated even some of his own supporters. In his books he used the foulest gutter language to vilify his opponents. He seems to have been paranoid and, like most paranoids, he developed real enemies. His self-praise is typical of certain schizophrenics. "Every little hair on my neck knows more than you and all your scribes. My beard has more experience than your high colleges," he wrote, and "Fortunate Germany that has me for her physician!" Still, he was capable of forming enduring friendships with men of intellect and discernment—Erasmus, for one.

Paracelsus's medical background was exceptionally broad. His theories were based on his experience with remedies of the school of John of Rupescissa; his training in both theoretical and practical alchemy; some studies at medical school; his work as a military surgeon and as an unlicensed medical practitioner; and extensive readings in medical classics, astrology, mysticism, and magic. But, his diatribes against physicians and the medical establishment were so violent that it seems apparent that his medical theories were not only the result of an intellectual synthesis, but also an emotional rejection of medical authority.

As a surgeon, he had been at the bottom of the medical caste system, and he openly resented and disliked the physicians at the top. Then, as one who had studied medicine but had not received the M.D. degree, he disliked the university system that had rejected him. Finally, Paracelsus, a fundamentalist Christian, could accept neither the pagan Galen as the authority, nor Galen's Four Humors theory as the basis of medical practice. And so he worked out his own medical system.

He began his theory of disease and treatment with what all Europe had known since the first terrified refugees fled from the Black Death, but medical theory ignored: that plague, syphilis, smallpox, and other diseases spread from person to person. Disease, therefore, was not just a condition of the body, but an invasion from the outside by some evil agent. As a practicing alchemist, he knew that metals as well as plants grew from seeds, so he reasoned that diseases also must grow from seeds. Perhaps he got these ideas from the Moors. In Moslem Spain in the 1400s, ibn al-Khatib had pointed out that people could be infected by the clothes and household objects of someone suffering from plague, by ships from an infected place, and by persons carrying the disease who were themselves immune to it (*6*). Paracelsus might also have picked up the suggestion from the works of Gentile da Foligno (d. 1348) or his own contemporary, Fracastoro, both of whom used the words "seeds of disease" for airborne particles. It is also quite possible that he thought of it himself.

In any case, the concept of seeds of disease led to a new and fruitful idea. Just as each species of plant came only from its own seed, so each disease came from its specific seed. Logically, because each disease had its own seed or cause, so each would have its own particular cure, a specific medicine. During his wanderings, Paracelsus had seen distilled materials used effectively as medications. He therefore reasoned that Rupescissa's distilled quintessences were indeed cures, each one being a specific, effective against one particular disease, but not a panacea for all diseases. This simple idea immediately involved him in conflict on all sides, with alchemists and physicians alike. The very idea that each disease had its own cure was a rejection of the concept of the elixir of life, the universal cure the alchemists sought. On the other hand, using distillates as medications meant rejecting the theoretical basis of orthodox Galenic medicine.

The Galenists had early recognized that there was a contradiction between their own definition of disease as an imbalance of humors and Rupescissa's concept of disease as a corruption to be combated by a distilled quintessence. They therefore rejected distilled medicines. Paracelsus, also recognizing the contradiction, accepted their logic but reversed their conclusion. Knowing from experience that some distilled medicines really were effective, he denied the validity of Galen's entire theory of disease and treatment. Instead, to justify his own methods, he developed a medical theory based on alchemy.

As an alchemist, he accepted Aristotle's Four Elements theory (although, as a devout Christian, in some of his writings he rejected fire because it is not mentioned in Genesis). He also accepted the Hellenistic idea that metals were composed of sulfur, the principle of combustibility representing the soul, and mercury, the principle of volatility representing the spirit or intelligence. He now made an important modification in alchemical theory. He added salt as the principle of inertness, representing the body, and extended the concept to include all solids, not just metals. Salt as the third principle was readily accepted by alchemists. After all, man has a body as well as soul and spirit, so all things in nature must have body as well as soul and spirit.

All solids were made of sulfur, mercury, and salt, consequently the human body was made of sulfur, mercury, and salt. Each organ of the body had its own chemical composition and its own specific function, being operated by a spiritual force or vital essence, called an *archeus*. The overall system of organs was controlled by a Grand Archeus, which regulated the activities of all of the subordinate archei. Disease occurred when a malevolent archeus entered the body from the outside and attacked one of the good archei. Each evil archeus attacked a specific good archeus and injured a specific organ. The remedy to be taken for each disease therefore had to contain a specific counter-archeus, found in a mineral or a plant.

The chemical remedies Paracelsus prescribed often contained heavy-metal salts that acted as bactericides and worked effectively against infections, sometimes actually curing the disease. Bismuth and antimony treatments for gastrointestinal disorders have already been discussed. Mercury salts were applied to skin conditions, including syphilitic lesions. (Oddly enough, Paracelsus made little use of antimony compounds.) Most of his remedies were found by trial and error and, if they turned out to be effective, some sort of rationalization was worked out. For example, anemia was found to respond to a diet containing iron salts. The explanation offered was that blood was red, the same color as the planet Mars, which was associated with iron, the metal from which armaments were made. It followed that pale blood resulted from a lack of iron.

Paracelsus was not simply an adventurous and imaginative healer with practical experience in alchemy. Like many natural philosophers of the Renaissance, he was a mystic who believed that in studying man one learned about nature and in studying nature one studied man. He believed he had a divine twofold mission: to heal souls through Scripture and to heal bodies through treatments based on astrology and alchemy.

His medicine was based on theology and philosophy as well. He admonished physicians, "None among you who is void of astronomical knowledge may be filled with medical knowledge." He urged also that physicians practice alchemy, and wrote that the true physicians

> ... do not consort with loafers or go about gorgeous in satins, silks, and velvet, gold rings on their fingers ... but they tend their work in the fire patiently day and night. They do not go promenading but seek their recreation in the laboratory, wear plain leathern dress and aprons upon which to wipe their hands, thrust their fingers among the coals into dirt and rubbish. They well know that words and chatter do not help the sick nor cure them ... therefore, they busy themselves with working with their fires and learning the steps of alchemy [4].

His *materia medica* was based on astronomy (which in those days also meant astrology). The planets in the sky corresponded to parts of the human body—the sun to the heart, the moon to the brain, Mercury to the lungs, Jupiter to the liver, Venus to the kidneys, and Mars to the gall bladder. To cure liver disease, one looked for minerals or plants corresponding with the planet Jupiter. Obviously, whatever material proved to be effective in treating liver problems was then considered, somehow, to be connected with Jupiter, and some explanation was concocted. Just searching for these correspondences involved a study of nature that greatly increased knowledge.

Materials that might correspond with the stars or planets were processed to obtain medications. The processing was typically alchemical in the tradition of John of Rupescissa. To Paracelsus, the important step in the production of a specific was either distillation or sublimation, because only the pure essences had curative powers. Unfortunately, he concentrated on the distillates and ignored the residues left in the distilling flasks that contained any bactericidal substances. For example, he dissolved each of the known metals in various acids. Then, he distilled each solution, saving the water and any excess volatile acid, which were essentially without therapeutic effect, but throwing away the solid residues.

Paracelsus defined alchemy as the art of transforming natural raw materials into finished products. To him, the cook, the baker, the weaver, and the apothecary were all alchemists, and he lamented that they had not studied sufficient alchemy to be as expert as they should be. Elsewhere he defined alchemy somewhat differently, as the art of transforming raw materials by fire. However defined, alchemy to him was more than just the mechanical operations performed on materials. It was the study of the cosmos, because the stars were connected to the chemicals studied by the alchemist.

Paracelsus always had friends and followers. However odd his actions might seem at times, or how difficult his personality became, people could not fail to be impressed with his sound medical common sense and his occasional spectacular cures. He really was a very good physician; and by the standards of the time, he was outstanding. Some of his remedies actually did help, and even those that did no good at least did no harm. His patients were able to overcome their diseases because their body resistance had not been weakened or destroyed by Galenic treatments. Even though he was famous, during his lifetime his ideas had little impact because his books had not yet been published. When, about twenty years after his death, they began to appear in print, people were eager to learn more of this strange man and his ideas. Consequently, although his writings were obscure, they reached a wide audience that was receptive and sympathetic.

The School of Paracelsus: Iatrochemists

The publication of Parcelsus's books shook the organized medicine of the day. Many physicians tried his treatments, and some were persuaded by the results. Among them were Joseph Duchesne (1544–1609), physician to Henry IV of France; Torquet de Mayerne (1573–1655), physician to Henry IV of France and James I and Charles I of England; Peter Severinus (1542–1602), physician to the King of Denmark; Oswald Croll (1580–1609), physician to the Prince of Anhalt–Bernburg; and Johann Hartmann (1568–1631), physician to the Landgrave of Hesse–Cassel.

Joseph Duchesne, like Paracelsus, refused to consider fire as an element because it was not mentioned in Genesis. He directed his attention to the chemicals found in natural mineral waters. If the spring waters at the various spas had any medicinal value at all, it must be because of the chemicals dissolved in them. When his analyses showed niter, alum, vitriol, sulfur, pitch, antimony, and lead in spa waters, he decided that these chemicals might have curative powers and should be studied. Duchesne, unlike his master, was a compromiser. Somehow he managed to accept Paracelsus's remedies at the same time as Galen's theories.

Johann Hartmann was the first regularly appointed professor of chemistry in Europe. He became professor of chymiatria, or medical chemistry, at Marburg, Germany, in 1609. (Chymiatria became iatrochemie or iatrochemistry, and later became chymia, chemie, and chemistry.) The appointment of a professor of chemistry at a medical school is perhaps the best indication of the importance that medical chemistry attained within fifty years of the first publication of the books of Paracelsus. Hartmann used antimony and mercury preparations, but he was not yet ready to make a clean break with the past, and so he also prescribed Galenic remedies. To cure insanity and melancholia, he prescribed blood drawn from behind the ears of an ass, emetics, holy water, powdered magnet, spirit of human brain, and the livers of live green frogs.

Oswald Croll, in 1609, wrote the *Basilica Chymia*, the first authoritative listing of the various materials and procedures of Paracelsus. He believed that a physician must not only be a chemist but also must have religious training. He was a university-trained physician who retained part of the traditional approach. He still prescribed mummia, which had been used for centuries, and which even Paracelsus prescribed. One recipe for it used the corpse of a twenty-four-year-old, red-headed man who had met a violent death. The body was cut into bits, spiced, soaked in alcohol, and then extracted with alcohol. If this was a remedy that even the followers of Paracelsus used, the Galenic remedies they rejected must have been vile indeed. Croll accepted Paracelsus's idea that sympathies, not antipathies, cure. In other words, if the patient has a liver problem, feed him liver. This idea probably came from folk medicine, based on sympathetic magic, and Croll recommended many treatments from folk medicine. Nevertheless, he also prescribed chemical medications including calomel, potassium sulfate, succinic acid, calcium acetate, and mercury and antimony compounds.

Theodore Torquet de Mayerne had practiced in Paris, where he was one of the physicians to Henry IV. He prescribed antimony, tin, mercury, and iron compounds. For these prescriptions, he was vilified, and the faculty of the University of Paris, an orthodox stronghold, had him condemned as a follower of Paracelsus. Ultimately he was forbidden to lecture and was forced to leave Paris in 1606 for England, where he became court physician to James I and his son Charles I. A cultivated man, he was interested not only in medicine and medical chemistry but also in the chemistry of colors. He worked with Van Dyck and Rubens to develop new paints and pigments.

It is significant that so many royal and ducal physicians were followers of Paracelsus. One would expect that only the established medical schools would have had strong connections with the royal courts, and that the court physicians would be the most eminent Galenist practitioners. Instead the number of court physicians who were iatrochemists is out of all proportion. The probable reason is that physicians who used the new single-ingredient medicines cured some of their patients, while others recovered by themselves. The death rate of their patients dropped sharply and their reputations spread rapidly, reaching the ears of local noblemen and through them, the royal courts. The relatively few followers of Paracelsus became influential, and the downfall of Galenic medicine was inevitable.

When the Paracelsian physicians began to concentrate on noble and wealthy patients, the public had to turn elsewhere for medical treatment. Rather than go to a Galenist physician, the peasant or the laborer or the shopkeeper would go to an apothecary[2], tell him the symptoms, and ask for a drug to cure the sickness. Because the apothecaries gave more effective medicines than the Galenic physicians, their public image improved and their status rose. In England in 1608, they separated themselves from the Company of Grocers, the original sellers of drugs, and formed their own group, the Company of Apothecaries. In 1703 they were given the right to practice medicine, a right that was withdrawn only in the nineteenth century.

Paracelsus's Contributions

Paracelsus's savage attack upon the Galenic treatments came just as the public, and perhaps even some physicians, were losing confidence in classical medicine. Galenic remedies had been totally ineffective against the plague, syphilis, smallpox, and all the other infectious diseases that were devastating contemporary Europe as a result of denser population and easier travel. Paracelsus's theory that disease had an external cause put medical practice on a much more rational basis. Physicians now began to search for cause-and-effect relationships. His preference for one-ingredient drugs for specific diseases gave physicians a method for determining the effectiveness of a remedy. Finally, his emphasis on carefully controlled dosage not only saved many lives, but also gave physicians an understanding of the narrow limits within which a drug may be beneficial, and beyond which it may be either ineffective or positively harmful. After Paracelsus, medicine progressed relatively rapidly.

As for chemistry, his direct contributions were fewer but still significant. He started a system of classification of materials and

[2]Even today, in an era of high costs for medical treatment, many people go to a pharmacist for help. Unfortunately, if the pharmacist can't help, by the time the patient gives up and goes to a physician, the disease is relatively far advanced.

articulated the Three Principles theory that was later incorporated into the Phlogiston theory (*see* Chapter XI). His indirect contributions were much more important. His emphasis on laboratory work and experimentation influenced succeeding generations of chemists. His inorganic remedies, although not original, caught the attention of a wide group of practitioners, and the resulting continuous search for new medications produced a flow of materials and reactions that forced the development of a system of classification.

Paracelsus's followers, the iatrochemical physicians, who were administering potent and potentially poisonous drugs, paid careful attention to the purity and the quantitative composition of their materials. As a result, they made rapid strides in chemical analysis. Paracelsus stressed that all techniques involving materials were part of the domain of alchemy, an idea that led soap boilers to find out what metallurgists were doing and that caused chemical knowledge to spread across intercraft barriers. He was the first great alchemist to rebel against the cramping, crippling authority of the Greeks.

Physicians now not only employed the services of medical alchemists, but also began to work in medical alchemy themselves. By the late sixteenth or early seventeenth century, the term *alchemy* became restricted to the operations of the gold-seeker, and the terms *chymia* and then *chymistry* were applied to the study of matter. Chemistry, at first a service branch of medicine, became a discipline in its own right. By 1661, the wealthy nobleman Robert Boyle would call himself a "skeptical chymist."

References

1. Multhauf, R. P. "John of Rupescissa and the Origin of Medical Chemistry," *Isis 1967, 45,* 359–367.
2. Garraty, J. A.; Gay, P. *The Columbia History of the World;* Harper and Row: New York, 1972; pp 326–328.
3. Fairbank, J. K.; Reischauer, E. O.; Craig, A. M. *East Asia;* Houghton Mifflin: Boston, 1963; pp 171–172.
4. Pagel, W. *Paracelsus;* Karger: Basel, Switzerland; 1958.
5. Partington, J. R. *History of Chemistry;* Macmillan: New York, 1961; Vol. II, pp 115–151.
6. Crombie, A. C. *Medieval and Early Modern Science;* Doubleday: Garden City, NY, 1959; Vol. I, p 229.

VIII

The Information Explosion

In medieval Europe, books were scarce and costly. Exquisitely illustrated and written in beautiful calligraphy on parchment by trained scribes, they were so expensive that only the very wealthy, such as princes, bishops, and bankers, could possess more than a few original manuscripts. In the schools and universities, most students did not own books, and reading was a cooperative effort. The entire class gathered around the lecturer (the word originally meant reader), who stood at a lectern as he read the text aloud to the group. The students wrote down what they heard and after the lecture they studied their notes. Sometimes students with more complete notes sold them to other students. Collections of such materials were passed along from generation to generation of students and, added to and amended, often ended up as books themselves.

Medieval Books

Books were duplicated by the same sort of procedure as that of the lectures. One person read the original aloud to a group of copyists, each of whom wrote down what he heard—or thought he heard. Inevitably each copy had its own spelling and punctuation, words were occasionally omitted, and errors inserted. Understandably, the end results, books copied from copies, were often very different from the originals.

Almost none of these expensive books was available to the rare literate artisan. The few books for craftsmen, such as the

Mappae Clavicula and the *Compositiones ad Tingenda*, were to be found only in monastery and cathedral libraries (*see* Chapter V). There they were available to monks and lay brothers involved with religious paintings and mosaics, parchment, inks and pigments, leather bindings for church documents, stained glass for windows, and viticulture and distilling for production of sacramental wines and liquors.

By and large, in medieval Europe, as in all societies before the sixteenth century, literacy was confined to noblemen, clerics, merchants, and physicians. There was still a gulf between the educated class and the artisans. The natural philosopher was usually isolated from mundane matters, except perhaps warfare. On the other hand, the illiterate artisan, unaware of matters beyond his own particular trade or skill, could not develop the inventions and improvements needed to raise the population out of its misery.

The revival of economic life that got under way in the thirteenth century resulted in an increased interest in worldly matters and with it an increased need for books and educated men. The Church had always needed administrators; and with more cathedrals, monasteries, convents, and universities, it now needed more of them. City-states and, later on, national states began to emerge, and the growing bureaucracies needed men who were not only literate but also trained in Roman law. The new world of international commerce and finance needed clerks. But more was involved in the growth of literacy and education than just utility and ambition. Prelates, princes, bankers, and merchants had always needed clerks, but by then they themselves had developed an insatiable appetite for learning. By the fifteenth century Europe had become avid for knowledge. When printing with movable type was invented, making relatively inexpensive mass-produced books available, the pent-up demand produced an enormous burst of publishing that can only be called an information explosion. The professional writer and professional publisher now made their appearances in European history.

In the first fifty years after the establishment of Johannes Gutenberg's printing house at Mainz in 1447, an estimated eight million copies (*1*) of forty thousand books (2) were published. This flood of information touched all areas of knowledge and reached all classes of society. From the start, printing was a relatively large-scale enterprise. As early as the 1450s, Gutenberg's printing house had a staff of about twenty-five men. By 1500, Anton Koberger's Nuremberg printing shop employed a hundred men at three dozen presses (3). Books now were plentiful and relatively cheap, so that only the poorest laborer could not afford a second- or third-hand book, and even he could borrow books from friendly artisans or merchants. Almost anyone could learn to read print, even those without formal education, and the self-educated scholar and the literate artisan began to appear[1].

The classics were printed immediately and for the first time knowledge of science, both ancient and contemporary, was widely disseminated. Starting in 1480, scientific and mathematical works were printed, including those of the medieval physicists Robert Grosseteste, Richard Swineshead, Edward Bradwardine, Jean Buridan, Nichole Oresme, and Albert of Saxony and of the mathematicians Leonard Fibonacci and Jordanus Nemorarius (4). The books of Albertus Magnus, Arnold of Villanova (in 1504 and 1532), Raymond Lull (in 1514), John of Rupescissa (in 1561), and Roger Bacon were published. So were the works of Hippocrates, Theophrastus, Lucretius, Celsus, Pliny, Galen, al-Razi, Avicenna, and, of course, Aristotle. Geber's works appeared in print in the fifteenth century and again in 1521, 1525, and 1531.

The First Science Books

Not only were the works of famous men reissued in print, but when printers became aware of the size of the market, they often commissioned new books. Authors began to write books specifically for the printing press, that is, to make money. These new publications made the latest developments available to the reading public, and, in a significant innovation, some were in vernacular tongues rather than in Latin. Artisans began to be familiar with scientific knowledge, and books on practical technology appeared. These were how-to manuals on mining, metallurgy, glassmaking, and the like. These books had a unifying effect on the scattered technologies because, for the first time, artisans in different crafts could see basic similarities in their various processes. Moreover, knowledge stimulated further learning. Scientists reading these books became interested in practical technology and began to apply scientific methods to technological problems.

[1]The Chinese worker remained largely illiterate, at least until the mid-twentieth century, even though printing had been invented in China. The Chinese written language is so complicated that long and intensive study is needed, which until very recently the laborer simply could not afford.

The first printed books concerned with chemicals were those intended for miners, smiths, assayers, and physicians. Assayers, located at or near mines and smelters, analyzed rock samples to see if the ores were worth processing for the various metals, especially gold, silver, and copper. At commercial centers and mints, they parted, that is, separated and purified, gold and silver. Human nature being what it is, assayers also kept people honest by checking the percentages of gold and silver in coins and jewelry[2].

Assayers were rational, skeptical, practical, no-nonsense workers, startlingly modern in outlook. They weighed and analyzed their materials quantitatively. Hundreds of years before the principle of conservation of mass was stated, assayers were acting and relying on the idea that you should get out what you put in. They kept their eyes strictly on their own practical objectives, with consequences both good and bad for the development of science. Their methods and materials were later applied to all aspects of chemistry, but until the start of the nineteenth century, assayers paid little attention to theory.

Books for Miners

About 1500, in the mining regions of Germany, *Ein Nutzliches Bergbuchlein* ("A Useful Little Book on Mining") and *Probierbuchlein* ("Little Book on Assaying") were printed and circulated extensively. The *Bergbuchlein,* the first printed book on mining geology, was written in 1505 probably by Ruhlein von Kalbe (d. 1523) (5). The *Probierbuchlein* gave clear and accurate instructions for assaying. These books, intended for the artisan and written in German, were clear and eminently practical, with no trace of secretiveness and no cryptic references designed to be understood by only a few initiates. The information they contained was not new; it had probably circulated throughout the mining districts for many years, at first orally and then in manuscript, before finally being printed.

In 1912, Herbert Hoover, a famous mining engineer who later became president of the United States, wrote that with the exception of references to atoms, twentieth century works on dry assaying were very much the same as the fifteenth century

[2]According to the famous "Eureka" story, in the third century B.C. Archimedes discovered the buoyancy principle while investigating cheating by a goldsmith. He realized that if two apparently golden objects had the same weight, the one of pure gold should have a smaller volume and displace less water than the one of gold mixed with silver. He thought of this while in his bath and ran stark naked down the street shouting, "Eureka!" ("I found it").

Probierbuchlein. A couple of sections from the *Probierbuchlein* give an idea of the flavor of the book. The directions for setting up a laboratory stress the use and care of analytical balances:

> First order a good and accurate Cologne or Nuremberg assay balance with a long beam which is adapted and proper to lift the silver button. Take care that you lift nothing heavy with it for by that the balance will be lamed and weigh false. [Obviously, by the fifteenth century and perhaps even earlier, there was such demand for assayer's equipment that in Nuremberg and Cologne craftsmen specialized in balances. There was also a market for books intended for specialists in assaying and mining, or the *Probierbuchlein* and the *Bergbuchlein* would not have been printed.] For a second [balance] you should have a one-way balance that is stronger, with which you weigh copper and ore for the assay hundredweight. For the third, you should have a balance for weighing added material and lead, which carries 23 or 24 lot. It must be quite strong so that a mark can be weighed on it [6]. [A lot was one thirty-second of a German pound, and a mark was half a pound.]

The directions for separating silver and gold are also clear and direct:

> To separate silver from gold, take one part of silver which contains gold, one part of Spiessflas [antimony sulfide], one part copper, one part lead, and fuse together in a crucible. When melted, pour into a crucible containing powdered sulfur and as soon as poured in, cover it with a soft clay so that the vapor cannot escape. Then, let it cool and you will find your gold in a regulus [mass of metal]. Place this on a dish and submit it to the blast [7].

To submit to the blast is to heat it in a stream of air. The volatile oxides produced are swept away, but the heavy gold and some powder of silver and copper sulfates are left. This method of sweeping oxidizable impurities away with a stream of heated air is widely used in the modern self-cleaning kitchen ovens. The directions continue:

> To reduce silver to a powder and again to silver, dissolve in nitric acid. Take the resulting powder and pour into impure water which is warm or salty, and the silver settles as a powder [silver chloride]. To make silver again from it, take the powder and place it on a cupel and add to it the powder from the residue of the aqua regia and add lead and subject it to the blast when there is enough lead so that it encloses the powder. Otherwise it would be blown away. Blast it with hot air until it flashes [7].

This procedure is somewhat like that which Angelus Sala later on used to demonstrate that Aristotle was wrong and that a metal did retain its identity in a compound (*see* Chapter X). Either the

Distilling nitric acid for separating gold and silver. (Reproduced with permission from the Bettmann Archive.)

assayers were not alert to the theoretical implications of their own operations, or the times were not yet ripe for an attack on Aristotle—or both.

Books for Metallurgists

The next important book published on subjects allied to chemistry was *De La Pirotechnia* ("Of Pyrotechnology"), a textbook of practical metallurgy by Vanoccio Biringuccio (1480–1538), first published in 1544 and written in Italian for the man of action, rather than the scholar. Biringuccio had little interest in theory but was skilled in metallurgy, casting cannon, making gunpowder, and, it is said, counterfeiting money. His book covers the entire field of metallurgy and also applies itself to producing fireworks and gunpowder, making alum and other useful salts, distillation, making sulfuric acid by burning sulfur under a glass bell in the presence of water (the beginnings of the chamber process), making gold leaf, drawing wires, and making pottery. He also mentioned a method of reprocessing waste ores at mines. Biringuccio flatly stated that he did not believe in alchemy, that there had never been any transmutations, and that there never would be. Typical of his keen observation was his report that when lead was heated in a fire, its weight increased by about eight percent. (It picks up oxygen.) Typical of his medieval frame of reference was his astonishment at this, because "the nature of fire is to consume substances," not to augment them. The report was perhaps the first written mention of this important, puzzling phenomenon that was central to the work of Lavoisier. (*See* chapter XII.)

The Sixteenth Century Classics

The extremely influential *De Re Metallica* ("Concerning Metallic Things") was written by Georgius Agricola (Georg Bauer, 1494–1555) in the year 1550, but not published until 1555. (Herbert Hoover and his wife, Lou Hoover, wrote the only English translation in 1912 [5].) Agricola was a classicist who became interested in medicine, earned a medical degree, and spent his last thirty years in medical practice. He wrote several books on medical subjects, including a new edition of Galen. He lived his entire life in mining districts and was deeply interested in geology, mining, and metallurgy. He wrote a treatise on the nature of fossils, and he is considered the father of mineralogy because

Example of industrial-scale distillation of mercury. (Reproduced with permission from the Bettmann Archive.)

of his descriptions of crystal form, cleavage, hardness, and other mineral properties. The interest aroused by *De Re Metallica* was so great that it was translated into German within two years of its publication. Agricola discussed mining methods, assaying ores, concentrating and smelting ores, cupellation, dissolving with lead and with mineral acids, and amalgamating with mercury. He also described the production of alum, vitriol, sulfur, bitumen, and glass. He was skeptical about alchemy but would not commit himself to a definite opinion on transmutation.

In 1574 the assayer Lazarus Ercker (1530–1594) published another practical handbook, *Beschreibung Allerfürnemisten Mineralischen Ertzt und Berckwercks Arten* (an approximate translation of this daunting title is "Treatise on Precious Mineral Ores and Mining"). He used Agricola as a source but essentially omitted the mining and stuck to the assaying. Ercker's book remained the basic text on the subject until about 1775, a matter of two centuries. Ercker, who rose to be Warden of the Mint to the Holy Roman Emperor and was ennobled, took as his (freely translated) motto "Test it before you praise it," the remark of an experienced assayer.

Medical alchemy also interested sixteenth century readers. Physicians by now were quite involved with alchemical procedures. Hieronymus Brunschwig (1430–1512), whose real name probably was Jerome Saler, was a physician who wrote on surgery and explicitly declared himself a follower of John of Rupescissa. As such, he was very much concerned with distillation and wrote two influential manuals: in 1500, *Distillierbuch*

("Distilling Book") and in 1512, *Liber de Arte Distillandi* ("Book of the Art of Distilling"). His purpose was to show how to separate active medical principles from the residues that he thought were inactive. These books went into at least fourteen editions and must certainly have influenced Paracelsus.

Girolamo Fracastoro (1483–1553) was a physician, astronomer, and poet. He wrote on logic, physics, astronomy, and medicine. In one of his poems, a shepherd named Sifilo contracts a venereal disease (the one the Italians called the French disease, the French called the Spanish disease, and so on). Somehow the name shifted from the shepherd to the disease, now known as syphilis. Fracastoro, who had been influenced by the works of Lucretius, maintained that all substances were composed of atoms. He was perhaps the first Renaissance scholar to take Lucretius's poem *De Rerum Natura* as a serious work of natural philosophy rather than as merely a literary work. In 1546 in his book *De Contagione* ("Concerning Contagion"), he suggested that "seeds" of disease that were carried in the air could transmit disease from person to person. (His seeds of disease might well have been Lucretius's *semina morbi.*) He also suggested that the basis of chemical reactions was "sympathy and antipathy" among the various chemical substances. The idea of sympathy and antipathy was already about two thousand years old, going back to Empedocles in the fifth century B.C. and lasting into the seventeenth and perhaps eighteenth centuries. Even today we explain solubilities with the aphorism "like dissolves like."

About 1565, the posthumous works of Paracelsus began appearing in print and were eagerly snapped up by those who had heard of him. They went through many editions and caused controversy in the medical world almost to the point of bloodshed. Then, in 1604, the *Triumph Wagen Antimonii* ("The Triumphal Chariot of Antimony") appeared, purporting to be a newly discovered work by a fifteenth century monk named Basil Valentine. It contained many of the observations and methods that were also found in Paracelsus's works. Paracelsus's reputation immediately suffered and he was called a plagiarizer and an imitator. However, today's experts are convinced that the "Chariot" was plagiarized from Paracelsus and that Basil Valentine never existed. Probably the book was written by a printer and salt manufacturer named Johan Thölde, who had published some of Paracelsus's works and was familiar with their content. He was also a chemist of sorts and had already written a book on salts. It is still a mystery just why Thölde, if indeed it was he, would write under a pseudonym, because his earlier writings were published under his own name. Perhaps he did not wish orthodox physicians to connect him with the invective and ridicule that the "Chariot" poured out on them. Perhaps he felt that the book might sell better under the name of a supposedly ancient source. In any event, "The Triumphal Chariot" was strongly influenced by Paracelsus, and even used the same style and foul language. It extols the virtues of antimony as a medication but reminds physicians that antimony salts can also be poisonous. It gives careful directions for making antimony medicaments. It also describes arsenic and arsenic compounds, vitriols, the mineral acids, and even some organic products, including ethyl chloride, acetic acid, and perhaps acetaldehyde. For a century, "The Triumphal Chariot" remained the definitive work on antimony.

Works of Palissy

Most of the books mentioned so far, from the *Bergbuchlein* to "The Triumphal Chariot," were simply compendia of previously acquired information. The authors were primarily compilers and not experimenters, laying no claim to originality.

The next important author was an innovator, reporting on his own work. Bernard Palissy, a Frenchman (c. 1510–1589), had little formal education. He was an outstanding example of the new breed of self-educated artisan with a broad background obtained from reading the printed books now available. Not surprisingly, Palissy had little faith in the accepted authorities. He stressed the importance of independent research and experimentation and insisted that theories should always be checked by observation. By trade he had been a painter of glass but, in 1540, when he saw a porcelain cup made in Italy, he resolved to make his own porcelain. It cost him everything he owned and re-duced him and his family to dire poverty, but after seventeen years of effort he was able to make Palissyware, an enameled pottery that had a great vogue. He became famous and wealthy, but his troubles were not over. A Protestant, he was imprisoned for his religious beliefs. At first he was too useful to the government to be kept in jail, and so he was sent to Paris, where in 1566 he was given a position in the Tuileries gardens. He lectured on natural history and chemistry from 1575 to 1584. Finally, in 1588, the aged Palissy was again sent to jail, to the Bastille, where he died a year later.

Palissy's lectures had a considerable influence on the next generation of French scientists. Francis Bacon probably attended some of them while on a visit to Paris as a very young man. Palissy was interested in minerals and had a collection of labeled specimens, one of the first such collections. He published books on pottery, the classification of salts, the use of chemistry in agriculture, and related topics. His attitude toward transmutation is amusing. He considered it false and said that he could easily show it to be so, but he did not want to bother stopping it:

> Let them go on, that saves them from greater vices, since they have the means to try these things. As to the physicians, in following alchemy, they will learn to know nature and that will be of service to them in the art and in doing it they will recognize the impossibility of the business [*8*].

Palissy made no original chemical discoveries, but his influence was great in directing men's minds toward experimentation and toward the utility of their labors. The French have always been influenced by literary style, and Palissy, in beautiful literary French, calmly and quietly warned against the consequences of an "imaginative theory" and the harm done "by the imagination of those who have no experience." This warning was especially effective coming from a man who had indubitably created a new type of ceramic entirely out of his own experience.

Two other important books of this period were the *Magiae Naturalis* ("Natural Magic") and *L'Arte Vetraria* ("The Art of Glassmaking"). *Magiae Naturalis,* written by Gianbattista della Porta (1535–1615), was first published in 1558. Porta was not an experimenter and his book was essentially a survey of information from the available literature, a compendium of everything under the sun, much true and much false. The final edition, published in 1584, ran to twenty volumes. It included as useful information a section on poisons and antidotes. Porta also published in 1608 a practical laboratory manual on distilling in which he described methods, apparatus, and applications. *L'Arte Vetraria,* published in 1612 by Antonio Neri, was the first manual on making glass. It was translated into English from the original Italian, reissued in 1679, and served as the basic text for a century after that.

The Sixteenth Century Scientific Community

So far, even though we have touched on only a small fraction of the sixteenth century authors, our list includes Brunschwig, Paracelsus, Fracastoro, Biringuccio, Agricola, Palissy, Ercker, Porta, Tholde, and Neri. This age was no longer the medieval era, when a man working in medical chemistry or as a chemical artisan had no one with whom to exchange ideas and no way to find out what was happening elsewhere or even what had been done during the past fifty years. In the late sixteenth century, communication was relatively quick. These men knew of each other, and sometimes knew each other personally. They could discuss their work with their peers. They used each other as sounding boards for ideas, conducted arguments by correspondence, and cited each other's works. For the first time, there existed a community of scientists groping toward cooperation.

By the last quarter of the sixteenth century, the posthumous works of Paracelsus had split physicians into two groups, pro- and anti-Paracelsus. The former were now called iatrochemists (medical chemists), a name attributable to Paracelsus himself. The iatrochemists concentrated on inorganic compounds as medicines, and this inevitably led their attention away from the distillates and toward the residues left in the distilling flasks. In the long run, investigation of these solid residues was the real beginning of synthetic chemistry, the preparation of new chemicals and the attempt to find out what they could be used for. In other words, by now chemicals themselves were being studied. In the short run, however, it led to total disorder and confusion on the part of iatrochemists.

The large number of inorganic solids that the iatrochemists produced and even used as drugs had no place in the conceptual scheme of either alchemy or iatrochemistry. Paracelsus's ideas had been based on medieval alchemy: all medical materials were supposed to be mixtures of inert impure dross and of active pure quintessences or elixirs. (The alchemists spoke of elixirs, the iatrochemists referred to quintessences.) The differences between the various substances were considered to depend only on the differing proportions of the pure and the dross. Originally the iatrochemist's method of preparation was to separate the pure from the dross by fire, distilling the material, catching the essence by condensing it, and discarding the undistilled residue as inert trash. By definition, any solid residue was useless junk and perhaps even pernicious. (It must be admitted that Paracelsus, who was no model of consistency, had in his book *Archidoxies* attempted to classify some solids as *magisteries* ["medicines"], but without being specific.) Now, however, the iatrochemists discovered that many of their distilled essences were largely water and acid, without any therapeutic effect, while the supposedly inert residues were often active powerful medications. It was just the reverse of what they had expected. In addition to opposition from the Galenist physicians, the iatrochemists now had to contend with the fact that their new findings contradicted the very basis of their theory.

Andreas Libavius

At this stage, Andreas Libavius, actually Andreas Libau (c. 1540–1616), attempted to introduce some order and to moderate the dispute between iatrochemists and Galenists. Libavius was originally a professor of history and poetry at the University of Jena. Later on, he earned a medical degree and became interested in iatrochemistry. He rejected the anti-Paracelsist, or Galenist, doctrines of medicine, probably because he knew too many chemical remedies that worked. On the other hand, he was repelled by the excess of Paracelsus's diatribes and criticized him sharply. "Filthy are the Paracelsian lies and blasphemies," he wrote, and "Paracelsus, as in many other matters he is stupid and uncertain, so also here he writes like a madman." Typical of his intermediate position, he used Paracelsus's chemicals and methods, but emphasized the residues, not the distillates, probably the first to do so.

Libavius considered chemistry as "the art of producing magisteries and of extracting pure essences by separating bodies from mixtures." This statement implies a recognition that separations could be affected by dissolving and extracting with acids and alkalies, not just by heating. To him, chemistry was not just iatrochemistry, and its purpose was not just to be of use to medicine. It was a separate discipline that overlapped medicine.

Libavius was a very learned man who quoted citations and references from all available sources, which did not make his works easy reading. Partington says, "He attempted to extract sense from alchemical authorities, a task which sometimes proved too much even for his painstaking efforts" (*9*). Interestingly enough, he still believed in the possibility of transmutation.

His major work, the book *Alchemia*, published first in 1597, was a comprehensive survey of the chemical knowledge of the time. It included chemical information that until then had been considered to be part of metallurgy, as well as material from iatrochemistry and industrial alchemy. As a professional pedagogue, Libavius was actually making a serious effort to organize and systematize the apparatus and operations of chemistry so that it could be taught as a separate discipline. He divided his book, and presumably the field, into two parts, *Encheria,* which consisted of the methods of operation, and *Chymia,* which was the description of chemical substances and their properties. The term encheria never did catch on. Chymia, however, became accepted as including all chemistry and alchemy except for transmutation. The term alchemy was retained only for activities involved in transmutation and the production of elixirs.

Libavius's theories did not have too much direct influence. They were unclear, and his writings were mostly in Latin, which many artisans could not read. However, his laboratory instructions, some of which were written in German, were clear and to the point. The poor circulation of his book may have been because it was too theoretical at a time when laboratory instructions were in demand. Nevertheless, his attempt to organize chemistry as a formal discipline foreshadowed the future. It is not a coincidence that, only twelve years after the *Alchemia* was first published, formal instruction in chemistry began at the University of Marburg.

Books for the Medical Chemist

By 1600, the medical chemist, or iatrochemist, was no longer an alchemist looking for the Philosopher's Stone or the elixir of life or for various spirits and essences with magical powers. Either he was a physician, or he worked with a physician, to make chemicals with which to treat disease. He wanted to know and to use the simplest, safest, and best way to make those materials. He wanted them to have definite recognizable properties, so that he could be sure he was giving the patient the right drug. He wanted his compounds to be pure so as not to poison the patient accidentally. Consequently he wanted books on chemistry that might or might not contain philosophy, but that would at least give clear, concise directions on how to proceed, how to use the apparatus, and, especially, how to name his materials and recognize their names in the literature he read. Moreover, he was not satisfied to prepare only known remedies, because new discoveries were constantly being made. He had to know the latest preparations and how to make his own new chemicals. In short, he wanted a laboratory manual.

In 1608 the first of these manuals, the *Basilica Chymia* (freely translated, "Structure of Chemistry") was published by the physician Oswald Croll. It was a listing of recipes for Paracelsian medicines and an exposition of Paracelsus's system. Among other chemical medications, it listed calomel and tartar. It never attained the popularity of the book by Jean Beguin.

Beguin, who died about 1620, thought along the same lines as Libavius, but he had a more practical bent. He was less philosophical, and neither pro- nor anti-Paracelsian. In his book *Tyrocinium Chymicum* ("Chemistry for Beginners") (*10*), first published in 1610, he defined chemistry as an art "which teaches the dissolution of natural mixed bodies and upon their dissolution, their coagulation to make medicines more agreeable, salubrious, and assured." In other words, the chief function of the chemist is to aid the physician. Elsewhere in that book he is even more specific: "Chemistry, not acquiescing in the knowledge and contemplation of mixed bodies, as a natural science, but having regard to the work or business . . . unto it is worthily created a place among the practical arts and disciplines." Because chemistry was not a science but an art, why should one bother with natural philosophy? He therefore organized chemistry on an operational basis in terms of distillation, solution, and coagulation (or, to use our term, precipitation). His book is a collection of clearly written recipes that can be followed. In discussing the principles of science, he contrasts the viewpoints of the physician, the physicist, and the chemist. Even to Beguin, chemists had their own separate discipline and viewpoint. Beguin's first edition was only seventy pages long, but each edition grew, until the final was about five hundred pages.

Both Beguin and Libavius were aware of each other, and may have been personally acquainted. Each modified subsequent editions of his work to comply with the criticisms made by the other. Other laboratory manuals were published later in the seventeenth century. The most popular of these was the *Cours de Chymia* ("Course of Chemistry") by Nicholas Lémery (1645–1715), published in France in 1675, translated into many languages, and used all over Europe.

In 1450, if a scientist were lucky, he might have access to the works of a dozen or so natural philosophers and might even be acquainted with a few of them. The artisan knew only what he had been taught about his trade during his apprenticeship. By 1650 the situation had changed. Scientists had access to scores of books in various fields touching directly or indirectly on their interests, and artisans, for the first time, were able to read instructions and explanations concerning their crafts. In 1450, if a new development occurred, perhaps a dozen or so people might learn of it within the next decade. By 1650 hundreds of people were avidly reading each new book. New information was spreading rapidly all over Europe.

References

1. Clapham, M. In *A History of Technology;* Singer, C.; Holmyard, E. J.; Hall, A. R.; Williams, T. I., Eds.; Oxford University Press: London, 1957; Vol. III, p 377.
2. Ibid., p 390.
3. Russell, F. *The World of Dürer;* Time-Life Books: New York, 1967; p 37.
4. Crombie, A. C. *Medieval and Early Modern Science;* Doubleday and Company: Garden City, NY, 1959; Vol. II, pp 111–113.
5. Agricola, G. *De Re Metallica;* Hoover, H. C.; Hoover, L. H., translators; Dover: New York, 1950; pp 610–611. Reprint of translation published in *The Mining Magazine,* London, 1912.
6. Stillman, J. M. *The Story of Alchemy and Early Chemistry;* Dover Publications: New York, 1960. Reprint of *The Story of Early Chemistry;* Appleton: New York, 1924; pp 303–304.
7. Ibid., pp 305–306.
8. Ibid., p 348.
9. Partington, J. R. *History of Chemistry;* Macmillan: New York, 1961; Vol. II, p 246.
10. Multhauf, R. P. In *Great Chemists;* Farber, E., Ed.; Interscience: New York and London, 1961; p 74.

IX

Changing the Frame of Reference

The Undermining of the Scientific Establishment

AT THE START OF THE SIXTEENTH CENTURY, European natural philosophers were confined within the framework of Greek philosophy as modified by Moslem and Christian theology: Man was here on earth to play his part in the drama of salvation, demonstrating the greatness and goodness of God. The earth was the center of the universe, as shown by common sense, revealed in Holy Scripture, and proved by Aristotle's physics with the aid of astronomy as elaborated by Ptolemy. The works of Aristotle, Ptolemy, and Galen represented the pinnacle of secular knowledge, which was, of course, infinitely inferior to the Divine truths of the Scriptures. To question Aristotle and Ptolemy was to run the risk of being charged with heresy, because the Church's earlier tolerant attitude had hardened in the fire of reformist attacks against its wealth and worldliness.

The intellectual world of western Europe was a closed system. Education was in the hands of the Church, and all universities were religious institutions. Even where the teaching was not done by clerics, it had to conform to the Church's position, or prompt action would be taken by the authorities, both ecclesiastical and secular. In special cases, where the charge was minor and the man involved was useful or politically well connected, heretical ideas might be overlooked. But, on the whole, Church and state cooperated to prevent any important challenge to authority. Even in the following century, the Parlement of Paris, the highest judicial body in France, resolved in 1624 that "no person should either hold or teach any doctrine opposed to Aristotle . . . on penalty of death."

The advent of Protestantism in the sixteenth century shattered Christian intellectual unity. England, Holland, Scandinavia, parts of Switzerland, and large portions of Germany and France broke away from Catholic control. In those Protestant areas, except for Geneva, not only were the various churches unable to enforce intellectual conformity, but basic Protestant doctrine required that each Christian find the truth for himself by studying the Scriptures. Literacy and intellectual individualism were not merely permitted, they were a religious obligation. Everywhere, in both Catholic and Protestant areas, the religious struggle stimulated intellectual ferment.

Of course, wherever any one church was politically powerful, scientists and philosophers could deviate from the established line only at their peril, as Galileo found out. The Italian philosopher Giordano Bruno was burned at the stake in Rome in 1600. Van Helmont, Bernard Palissy, Nicholas Lemery, Girolamo Cardano, and Tomasso Campanella all suffered in various degrees from attempted thought control by Church authorities. The Protestants, too, persecuted dissidents. Jean Calvin burned philosopher-physician Michael Servetus at the stake in Geneva in 1553. Even in relatively tolerant England, although Roman Catholics were not maltreated, they were excluded from universities and from government positions until 1829.

Paradoxically, the first and perhaps the most important blow at Greek scientific authority was struck from within the Church. Copernicus was canon of the cathedral at Frombork (*Frauenburg*, in German) in Poland. His orthodoxy and devotion to the Church were beyond question; moreover, his theory that the earth and planets revolved around the sun was not at first considered significant. Even so, it was published only when he was on his deathbed, and Osiander, who wrote the introduction, felt it prudent to put in a disclaimer stating that the work was to be considered only as a mathematical fiction without physical significance.

When Copernicus began work, the accepted view of astronomy had been developed by the Alexandrian Claudius Ptolemaeus, known to us as Ptolemy (85–158 A.D.): The universe was a set of concentric crystalline spheres with the earth in the center and the abode of God and His elect in the outermost spheres, the Empyrean Heaven. All the spheres except the Empyrean Heaven rotated daily around the earth, carrying the planets, stars, sun, and moon with them. The stars were all fixed in one sphere so that they moved together, but the sun, the moon, and each planet had its own separate sphere, each moving in a different circular path. Circular motion was perfect motion, so all heavenly bodies moved in circles. Yet, it was easy to see that the planets wandered erratically, not in circles, sometimes even going backward. (Planet in the original Greek means "wanderer.")

To rationalize this contradiction, Ptolemy proposed that each planetary orbit was a combination of two kinds of circles, the epicycle and the deferent. The epicycle was a small circular orbit around which the planet moved. The center of the epicycle, in turn, moved around another, larger epicycle or around the deferent, a very large circular orbit centered on the earth. Eventually, to make the theory agree with observations, additional epicycles, up to eighty or more, were added.

Copernicus

Nicholas Copernicus (1473–1543), the Polish astronomer, rejected this complicated picture for typically medieval theological reasons. To him, the sun was the giver of light, and God would want the light to be in the center of His temple. In Copernicus's system, the moon continued to move around the earth, but the earth and all the planets moved around an equant, a point in space near the sun, each in circular orbits. The stars only appeared to be moving around the earth because the earth rotated on its own axis.

Today the idea of a heliocentric solar system is so familiar that we fail to understand why Copernicus's ideas were not immediately accepted. But for the sixteenth century natural philosophers there were good reasons for ignoring them. To calculate the positions of the planets with the Copernican system was almost as difficult and the results little more accurate than with Ptolemy's system. More important, however, the idea that the sun did not move not only contradicted Scripture (Joshua had commanded the sun, not the earth, to stand still) but completely impugned Aristotle's physics. To Aristotle, an apple fell because it was heavy and therefore moved toward the center, which, in Ptolemy's and Aristotle's astronomy, was the earth. But, according to Copernicus, the center of the universe was the sun. On the basis of Aristotle's physics, therefore, the apple should fall upward toward the sun, which obviously doesn't happen. If Copernicus was right about the center of the universe, Aristotle was demonstrably wrong about gravity. Then why did the apple fall? (Had Isaac Newton been born one hundred years earlier and believed in Ptolemy's astronomy, he never would have wondered why an apple fell.) There was nothing to replace Aristotle's physics, so natural philosophers gave little credence to the theory of the obscure Polish astronomer.

Unexpectedly, however, two remarkable events occurred. In 1572 a new and very bright star (a nova) suddenly appeared in the heavens, was visible for more than two years, and then disappeared. This event was a shock, not only to astronomers, but to all educated men. The starry heavens were not immutable. They did change. A fundamental belief, unchallenged for two thousand years, was suddenly, obviously wrong. Then, only three years later, in 1577 a comet appeared and moved through the upper heavens, where comets were not supposed to enter, and passed through Ptolemy's perfect crystalline spheres as if they were empty space. At this point, some astronomers began to develop serious doubts about Ptolemy's system.

Johann Kepler

The final rejection of Ptolemy's astronomy was due primarily to Johann Kepler, Galileo Galilei, and Isaac Newton. Kepler (1571–1630), a German, was a mathematical genius and a mystic. He worked with the data bequeathed to him by the Dane, Tycho Brahe (1546–1601), perhaps the greatest astronomical observer of all time. Tycho's data, taken without a telescope[1], were accurate to 2 minutes of arc, that is, 2/60 of 1 degree. Using this material, Kepler spent almost thirty years in miserable poverty, doing mind-breaking calculations without logarithms, tables of trigonometric data, or any of the other aids available to later mathematicians.

Originally, Kepler had set out to prove that the earth and the planets moved around the sun in circular orbits. After years of effort, he found that Tycho's observations differed from those predicted by the Copernican theory by only 8 minutes of arc; that is, about 0.04 percent. But Kepler knew that this minute discrepancy, small as it was, was still four times greater than the error of

[1]The telescope was invented nine years after Tycho's death.

Tycho's measurements. It had to be due to some defect in Copernicus's theory.

Kepler thought matters over very carefully and finally decided that the orbits could not be circular. This was an amazing decision. He had set out to prove that Copernicus was right, and after years of effort he had achieved almost complete success. Basing his calculations on the circular orbits postulated by the Church, Copernicus, Aristotle, Ptolemy, and all the astronomers who had ever lived, he had shown that Copernicus was 99.96 percent right. But to Kepler, 99.96 percent right was 100 percent wrong. With astonishing intellectual honesty and daring, he broke with everyone—Copernicus, as well as the Greeks and the Church—and cast aside circular orbits.

More years of trial-and-error calculations followed, while he eked out a bare living by casting horoscopes. Finally he settled on elliptical orbits and published his results. It had taken almost thirty years, but he had proved that Ptolemy was wrong and that even Copernicus had been only partially correct. The earth and the planets did go around the sun, but the orbits were elliptical, not circular.

Kepler's findings had little immediate impact. They were rejected by everybody, Ptolemaic and Copernican astronomers alike. Even his friend Galileo could not accept elliptical motion[2]. But Kepler now used his new system to calculate a set of astronomical tables that were invaluable to navigators. Published in 1627, they were dedicated to the Holy Roman Emperor Rudolph II in the vain hope that he would give Kepler some desperately needed money. The Rudolphine tables were so accurate that they were used all over Europe, even by those who rejected their theoretical basis and still clung to Ptolemy's system. They were simply too useful to be ignored, and as a result, the Copernicus–Kepler system constantly gained new adherents.

Galileo

A further attack on Greek astronomy came from Galileo Galilei (1564–1642), a professor at the University of Padua. Galileo had an almost completely modern viewpoint. An experimentalist, he used the inductive method and tried to place the laws of physics on a mathematical basis. For years a convinced Coper-

[2]Had Galileo been able to break loose from the medieval world and to accept the idea that natural motion was not necessarily circular, he would have arrived at Newton's First Law of Motion. Ironically, Kepler, the mystic, could dispense with uniform circular motion while Galileo, the pragmatist, could not.

nican, Galileo used the recently invented telescope to look at the sun and the moon. Through it he saw immediately that the moon had mountains and craters and was not the perfect sphere that Aristotle had supposed. Even the sun had moving, variable spots, so that the heavenly bodies were neither perfect nor unchanging. He saw, too, that Jupiter had moons, a fact indicating that bodies other than the earth had satellites. (He called Jupiter's moons the Medicean Stars for obvious political and financial reasons.) Unfortunately for him, Galileo did not confine himself to astronomy. He strongly attacked Aristotle's physics, although not Aristotle himself. (He maintained that if Aristotle had been able to look through the telescope, he too would have been a Copernican.) He urged his critics to see for themselves. When they refused, on grounds that there were too many well known optical illusions for them to bother with the telescope, he wrote to Kepler, "Tell me, should we laugh or should we cry?" In the final event, he cried.

The Church might have been able to turn a blind eye to Copernicus and Kepler, but it could not keep quiet in the face of Galileo's determined campaign, and after a period of toleration, it took action. In 1615, Galileo was forced by the Inquisition to abjure the Copernican theory, and in 1616, Copernicus's ideas were condemned, his books were banned, and it was forbidden to teach his theories. By then, however, it was too late. The books and theories were in circulation all over Protestant Europe, and they were being discussed at universities and taught to the next generation of astronomers. The final triumph of Copernicus and Kepler came in 1665–1666 when Newton used Kepler's ellipses as the basis of his gravitational theory and laws of celestial motion. With that, Ptolemy's astronomy and Aristotle's celestial physics became of historical interest only.

Other Assaults on Greek Science

Galen, too, came under attack. In 1543, the same year that Copernicus's book was published, Andreas Vesalius, a Flemish anatomist, published a remarkable book on human anatomy called *De Humani Corporis Fabrica*, showing that Galen had made mistakes in anatomy and physiology. Shortly thereafter, the books of Paracelsus began to reach natural philosophers and physicians with further attacks on Galen and, by implication, on the authority of Greek science.

A different kind of assault was delivered by Francis Bacon, Lord Verulam (1561–1626). Paracelsus, Copernicus, Kepler, Galileo, and Vesalius had denied specific ideas in astronomy, physics, and medicine. Bacon, however, called into question the basic concepts of Aristotelian science, the Greek scientific method, and the very aims and objectives of Hellenic science.

Bacon was not himself a scientist but a philosopher and a politician who at one time was Lord Chancellor of England. As a practical man of affairs[3], he was convinced that science should be useful, not just an auxiliary to theology. He wrote, "The true and lawful goal of the science is none other than this, that human life should be endowed with new inventions and powers." He believed that knowledge is worthwhile for its own sake, but that just because a scientific discovery is useful does not make it vulgar and beneath the notice of the philosopher–scientist. In fact, to Bacon, if a discovery was useless, that meant it had no contact with experience and reality and therefore was probably false. Scientists, he argued, should have close contacts with the crafts and trades. In line with this idea, he urged the King of England to establish a national scientific institution that would advance science and industry. The British did not adopt his program, but in 1666, J-B. Colbert, Louis XIV's Minister of Finance, established the Paris Academy of Sciences along lines similar to those suggested by Bacon.

Although there were many glaring inconsistencies in his ideas, Bacon emphasized the inductive method. He urged that all science be based on facts that could be used to develop hypotheses. He insisted that all hypotheses and theories be verified by experiment, and persistently attacked Greek philosophy for its aims and methods. It would be a gross overstatement to say that Bacon's works were definitive and that he changed the basic frame of reference of scientists. Undeniably, however, he had considerable influence in both England and France. The very fact that his tracts on such topics were widely read shows that there was conscious dissatisfaction with classical science.

In 1643 Evangelista Torricelli (1608–1647), Galileo's assistant, invented the mercury barometer, based on the vacuum that Aristotle had argued could not exist. In 1656, Otto von Guericke invented the air pump (the prototype of the vacuum pump) and

[3]Bacon was perhaps too practical. When imprisoned for taking bribes, he defended himself by explaining that although he did accept bribes he did not let them influence his judgment. In other words, he cheated the bribers.

used it in spectacular widely publicized demonstrations, showing not only that a vacuum could be produced mechanically, but that it might actually be useful[4].

Even before that, however, René Descartes (1596–1650), who still rejected the possibility that a vacuum could exist, had proposed a new mathematical method for science and a new view of the physical world. The publication and wide circulation of Descartes's *Discours de la Method* ("Discourse on Method") and his *Principia Philosophae* ("Principles of Philosophy") clearly showed that the medieval–Renaissance frame of reference was no longer acceptable to scientists.

One revolutionary feature of the new scientific era was the development of scientific societies, groups of natural philosophers with similar attitudes and interests. The earliest such groups began as a few interested people meeting from time to time at someone's home and discussing natural philosophy. Eventually these conversations among dinner companions became formalized into regular meetings of an organization. Predictably, however, such activities were suspect, and unless the group was politically well connected, the authorities took the earliest opportunity to suppress it. The first such groups were, therefore, short-lived. Nevertheless, they spread all over Europe. The *Academia Secretorum Naturae* ("Academy of the Secrets of Nature") met in Naples during the 1560s, but it was soon charged with witchcraft and dispersed. In Rome the *Academia dei Lincei* ("Academy of the Lynxes," meaning of the clear-sighted) met between 1601 and 1630. In Florence the *Academia del Cimento* ("Academy of Experiment") existed between 1657 and 1667. The *Societas Ereunetica* ("Society of Investigators") existed from 1622 to 1657 at Rostock on the Baltic. In Germany two groups were organized, the *Collegium Naturae Curiosorum* ("College of Curiosities of Nature") in 1652 at Nuremberg and the *Collegium Curiosum Sive Experimentale* ("College of Curiosities or Experiments") at Altdorf in 1672. In England the Royal Society for the Improvement of Natural Knowledge was founded in 1660.

At the end of the seventeenth century, the governments' attitudes toward science had changed. The authorities had awakened to the fact that science was useful. Scientific societies began to receive government backing. In 1700 the Berlin Academy was established by the Prussian government, and the Russian government set up the St. Petersburg Academy in 1724. By

[4]Von Guericke's demonstration that a vacuum could be used to permit atmospheric pressure to do work led within fifty years to steam engines, for pumping water out of mines. Coincidentally, Torricelli's work had originated with a request from the Archduke of Tuscany to Galileo for advice on pumping water out of flooded mines.

then, science was an entity in its own right—international, organized, free of religious domination, and engaged in spreading the news about the latest discoveries. Moreover, applied science was coming into its own as governments and industrialists began turning to the natural philosophers for help.

X

The First Chemists

Sala, Van Helmont, Boyle

In 1500 many people were working with chemicals, including physicians, medical alchemists, assayers, artisans, and natural philosophers. Had some seer revealed to them today's ideas of chemical composition and reaction, they would have immediately rejected them—and for good reasons. A hundred elements instead of four would have been much too complicated to grasp, especially since each of our elements consists of many other, smaller, particles. Studies that didn't explain why a chemical had certain visible properties and behavior, but instead focused on quantitative aspects of unseen entities, such as the angles and distances between invisible atoms, would have been incomprehensible. Yet, before the end of the sixteenth century, chemistry had come into being—not modern chemistry, but chemistry, nevertheless.

Throughout the sixteenth and seventeenth centuries, while Copernicus was challenging the authority of Aristotle and Ptolemy, while Paracelsus and Vesalius were challenging the authority of Galen, a new viewpoint was quietly emerging in natural philosophy: that of the chemist. For hundreds of years medical alchemists, iatrochemists, and assayers had been isolating, purifying, and characterizing new chemicals and discovering new reactions. Printed books had made this information available to those concerned with chemicals and their uses.

Hundreds, and perhaps thousands, of interested and informed people were by then aware of the new chemicals and reactions and of deficiencies in the old system. At first, new suggestions had been offered only with considerable trepidation. Later, with the weakening of the authority of Aristotle and the simultaneous decline in the power of the Church to enforce conformity, significant changes were proposed.

Any new developments could neither be radical nor occur rapidly. When confronted with new materials and new reactions, those trained in traditional theory would continue to think along old lines, but would accept some ad hoc suggestions to rationalize or explain new facts. These suggestions would modify the accepted views somewhat. The next generation, growing up with the slightly modified theory as its frame of reference, would be confronted with more information, demanding additional explanation. Each new suggestion, in turn, would change the theory still further. So each successive generation developed more modifications in its basic viewpoint. Slowly, the chemist's viewpoint changed from the classical to some sort of hybrid.

Jean Beguin's Compromise

One such example of a hybrid viewpoint is that of Jean Beguin (c. 1550–1620). He was typical of the expert laboratory technicians working with materials such as benzoic acid, acetone, and ammonium sulfate, all unknown to their medieval predecessors, and with scores of facts and observations that they could not fit into the old framework. He also was typical of the new breed of chemical philosophers, deeply interested in chemicals as objects worthy of study while still believing in the old theory. In 1612 he seemed to have arrived at the idea that chemicals are constituted of other chemicals, not of the Four Elements or the Three Principles. In the 1612 edition of his book *Tyrocenium Chymicum* he reported that he had dissolved minium (red lead oxide) in vinegar, digested it for a month, and then distilled it until the residue decomposed to a sweet-smelling distillate that he called "Burning Spirit of Saturn." Lead was believed to correspond with the planet Saturn, so he obviously believed his distillate contained lead or was derived from lead. He implicitly was contradicting Aristotle by suggesting that lead existed in the minium and perhaps in the distillate. Later, in the 1615 edition, written in French and with the name changed to *Elemens de Chimie* ("Elements of Chemistry"), he was explicit (*1*). He reported that when corrosive sublimate (mercuric chloride) and stibnite (antimony trisulfide) were heated together, the mercury in the sublimate combined with the sulfur from the stibnite, leaving the antimony to combine with the "vitriolic spirit" (chlorine) from the sublimate. His products were cinnabar (mercuric sulfide) and butter of antimony (antimony trichloride). The reaction, expressed in Beguin's terms, is

$$\text{stibnite} + \text{corrosive sublimate} \longrightarrow \text{butter of antimony} + \text{cinnabar}$$

In our symbols it is

$$Sb_2S_3 + 3HgCl_2 \longrightarrow 2SbCl_3 + 3HgS$$

Looking at the equation, in whatever symbols or terms, it is obvious that Beguin knew that cinnabar was composed of mercury and sulfur, butter of antimony was made of antimony and chlorine, stibnite consisted of antimony and sulfur, and corrosive sublimate was composed of mercury and chlorine. He realized that when chemicals combine they do not lose their individual identities but continue to exist in the compound.

Beguin actually had described and given a correct explanation of what we now call a double decomposition or double displacement. But he never stated explicitly the general conclusion that chemicals were composed of other chemicals. Educated in the sixteenth century, he thought along traditional lines despite the implications of his own work. Sooner or later, however, as more new information was accumulated and absorbed, existing ideas would be modified, and a new generation of chemists would break with many of the old ideas.

The Revival of Atomistic Ideas

Quite early in this long process of changing fundamental points of view, a fruitful stimulus came from an unlikely source. During the medieval era, Lucretius's 57 B.C. poem *De Rerum Natura* ("On the Nature of Things"), expounding the atomic theory of Epicouros, was known from references to it and from some quotations from it. The full text was not available, and few scholars paid attention to this work by a minor Roman poet. Then in 1417, during the revival of interest in classical culture, a complete copy was discovered. It was printed in 1473, and attitudes began to change. At first only the humanists took notice, but the idea that matter was composed of atoms became more and more attractive and convincing, especially to the iatrochemists and to those philosophers who were aware of the work of the Arab atomists. By 1600 there were some thirty editions of *De Rerum Natura* in all the major European languages. Ultimately, the basic idea—and the word *atom*—carried over into modern science.

The first, partial, rejection of Aristotle's chemical theory and acceptance of the idea of atoms, of which we know, came from Girolamo Cardano (1501–1576), the picaresque mathematician and physician. About 1556, he decided that fire was not an element, and anticipating Francis Bacon by some fifty years, he wrote that fire was a mode of motion of atoms. The atomist viewpoint spread rapidly, and in France, Jean Bodin (1530–1596) wrote of atoms of chemical substances in a work published in 1596. A further modification of the accepted theory came from the Dutch student David van Goorle (1591–1612). In a book published shortly after his untimely death, he suggested that neither fire nor air is an element and that water and earth exist in the form of atoms.

Other reformers and revolutionaries were also at work. The priest Tomasso Campanella (1568–1639)[1] completely rejected the Four Elements theory and the Three Principles theory, although he still believed in transmutation, which was based on the Four Elements theory. About 1620, Sebastian Basso, a French physician, proposed another atomic theory, in which every substance was composed of small characteristic particles or atoms, each with its own specific nature and unlike any other kind of particle. This was a further departure from the Greek belief in simplicity, the idea that only a few fundamental elements make up innumerable substances. In 1619, Daniel Sennert (1572–1637), professor of medicine at the University of Wittenberg, proposed a hybrid theory, with each of the Four Elements and Three Principles consisting of atoms of some kind. He used the term *phlogiston* for philosophic sulfur, the principle of combustibility, although he was not the first to do so. In 1606 his older contemporary, Hapelius (1559–1626) already had used the term *phlogiston*, and it is probable that Sennert picked it up from him. In 1605, Francis Bacon wrote favorably of Democritos's atomic theory, although his ideas changed from time to time. Eventually he decided that the properties of materials could be explained in terms of the size and motion of the particles of which they are composed, although the particles themselves might not necessarily be atoms. After Bacon, most English scientists were atomists.

---

[1]A surprising number of these early atomists got into trouble with the authorities. Cardano was in constant difficulties with the Inquisition. Campanella spent twenty-seven years in the prisons of the Inquisition, and van Helmont (discussed later in this chapter) spent ten years under house arrest.

At first, the revived atomic theory had few adherents. The majority of iatrochemists, natural philosophers, and technicians probably paid no attention to these new ideas, being, as we are, desirous of holding on to the intellectual capital acquired with so much difficulty. Nevertheless, the dates of the first modifications of the Four Elements theory and the early mentions of atoms range from about 1550 to 1625 and indicate a growing tendency to think in terms of atoms. Automatically, those who thought chemicals were made of atoms began to think about the size, shape, and weight of particles or atoms; what they were made of; how they were arranged; and the relation between the properties of atoms and those of macroscopic bodies. They were now thinking in the way chemists think and about the things that concern chemists, and in a short time they were writing and theorizing about these subjects. We therefore date the beginnings of chemistry to the last few years of the sixteenth century.

Angelus Sala

The first definite appearance of the chemist's approach came between 1617 and 1618 in works published by the great and almost forgotten Angelus (or Angelo) Sala (1576–1637). He was an Italian Calvinist who, as might be expected, was receptive to anti-Aristotelean ideas. He was an iatrochemical physician and, as a follower of Paracelsus, rejected the Four Elements theory in favor of the Three Principles theory. He was deeply interested in chemicals, their properties, their reactions, and their internal structure.

In 1617 and 1618 Sala reported on a series of experiments showing that chemical compounds were composed of other chemicals that continued to exist in the final products. In Moorish Spain and in seventeenth century Hungary, water containing copper sulfate was passed over pieces of iron to obtain metallic copper. (Pliny had reported this procedure some time before 49 A.D.) It was widely considered that somehow the iron was transmuted to copper. Libavius, for example, in 1596, wrote that the "vitriol water" (sulfuric acid) transmuted the iron into copper. Sala, however, demonstrated that there was a whole series of such reactions in which iron removed copper from solution, copper removed silver, and mercury removed both silver and gold[2]. These results indicated that the iron–copper vitriol reaction was not really a transmutation, but only an ordinary chemi-

[2]Ercker, in 1574, had reported that copper precipitates silver from solution just as iron precipitates copper (2).

cal reaction, and that copper existed in the blue vitriol solution not as earth or water but as copper. Not only did Sala show by this that traditional thinking was wrong, that a metal continued to exist in its compounds, but he actually pointed out that the iron–copper vitriol reaction was what we call a displacement, and that an entire series of chemicals reacts in the same manner.

Sala thought that the iron attracted the copper from the solution, leaving the vitriolic acid behind. But even though he was wrong on particulars, he was correct in thinking that chemical reactions such as this were rearrangements of internal particles.

In his discussion of fermentation, he was even more explicit, defining fermentation as the interior motion of the particles of bodies, resulting in the formation of new arrangements (3). In other words, he considered that substances were composed of combinations of particles and that the new arrangements meant new substances.

In 1617, in the third edition of his book *Anatomia Vitrioli* ("The Anatomy of Vitriol"), Sala reported on an extraordinary experiment demonstrating this new viewpoint. He planned and carried out a series of operations designed to test a point of theory of the structure of chemicals. He weighed some copper and dissolved it in hot concentrated sulfuric acid, added water, and obtained blue vitriol (copper sulfate hydrate). First he showed that this product was identical with natural blue vitriol. Then he converted it to copper oxide. Finally, he reduced the copper oxide back to copper, probably with charcoal, weighed the reduced copper, and compared it with the weight he had started with. Not only had he recovered the original copper, but within his experimental error the weight of the recovered copper was the same as the weight with which he started. The sequence was

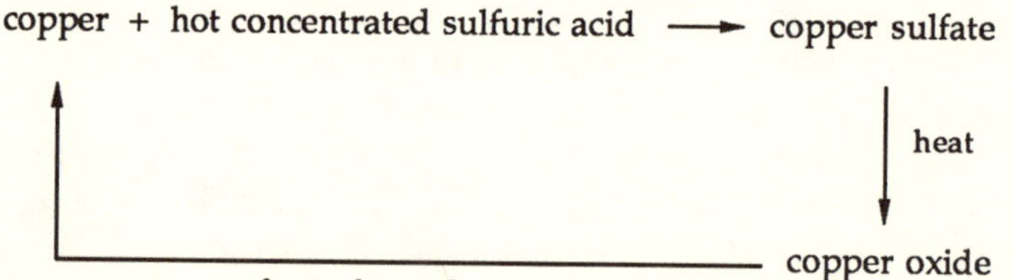

Obviously, such a complicated three-step process was performed only to prove or disprove a point. Sala had actually planned an experiment to demonstrate that copper maintained its identity in copper sulfate. Not only had he proved the Four Elements theory wrong, but, in fact, he had also proved Paracelsus wrong: Copper sulfate was composed of copper and sulfur, not of either the Three Principles or the Four Elements. We do not know how he rationalized his continued belief in the Three Principles with the results of his own experiment. Perhaps he ignored or tried to ignore the contradictions. Perhaps he did not fully realize the implications of his own work.

Sala was definitely a chemist in our sense of the word. His attitude toward transmutation was a century ahead of his own time. He not only attacked the possibility of transmutation but asserted flatly that the fundamental idea was wrong.

His demonstration that chemicals consisted of other chemicals refuted the idea that the addition of qualities or principles could change one substance into another. Moreover, when he showed that his synthetic blue vitriol was identical with natural blue vitriol, he dealt a double blow to Greek theory. Not only had Aristotle and his successors believed that minerals grow and ripen in the ground, but in fact this growth and ripening process was exactly what the alchemists were attempting to duplicate in their transmutation procedures. If a mineral such as blue vitriol could be synthesized quickly and completely in an ordinary chemical reaction performed in a laboratory, it could not really be a living, growing entity with a soul. The failure to transmute metals was not just due to technical difficulties. The basic premise of transmutation was wrong.

It is tempting to say that Sala's work was revolutionary, but in fact his ideas and their implications had little effect on his contemporaries. As late as 1738, Herman Boerhaave, a physician, chemist, and author of a widely used textbook of chemistry, still believed that minerals and metals had some sort of life. But, the idea that metals and minerals could be grown by some sort of alchemical process had lost credence, and even those, like Boyle, who still believed in transmutation, now thought in terms of the rearrangement of the internal structure of atoms.

Johan Baptista Van Helmont

A more important and better known figure than Sala was Johan Baptista van Helmont (1579–1644). Like Sala, van Helmont was primarily an iatrochemical physician, probably the greatest of that school, and he is more significant in the history of medicine than of chemistry. However, he is credited with the discovery of gases, which was central to the further development of chemistry. Most of the fundamental work of Joseph Black, Antoine Lavoisier, and John Dalton that established modern chemistry was done with reactions involving gases. Partly this was possible because gases could be easily prepared in a pure condition and partly because the arithmetical relationships between the volumes of gaseous reactants and products are simple and general.

Van Helmont had a great reputation both as a chemist and a physician. A wealthy Flemish nobleman and a devout Catholic, he practiced medicine only as charity. In spite of his philanthropy and his blameless personal life, he ran into trouble with the Spanish Inquisition (Belgium was a Spanish possession at that time) ostensibly for his comments on the medical value of saints' bones, but probably because his scientific radicalism made his ideas suspect. He was at first imprisoned and then kept under house arrest for the last ten years of his life. An odd mixture of the old and the new, as might be expected during a transition era, he was more medieval than modern. Like his younger contemporary, Robert Boyle, he was clearheaded but credulous. In medicine, he was an enthusiastic supporter of Paracelsus, rejecting sweating and bleeding, remedies that nonetheless lasted into the nineteenth century and perhaps beyond. (Mark Twain remarked that about 1850 his family doctor didn't allow blood to accumulate in the system.) On the other hand, van Helmont still dosed himself and his patients with foul remedies, such as worms from the eyes of toads. But he was a skilled and scrupulous observer who relied extensively on the balance and who believed that nothing could be wholly destroyed in a chemical reaction.

Van Helmont explicitly rejected the Four Elements theory for religious reasons, partly because Aristotle was a heathen and partly because Aristotle's theory was in contradiction to his own interpretation of the Scriptures. In Genesis, water is described as being in the heavens, and fire is not mentioned. Fire was therefore not an element or even a form of matter. Water and air were the only elements and neither one could be made of the other. Air did not take part in chemical reactions, and therefore all liquids and solids must be formed from water. (This is a variation of another pre-Christian theory, that of Thales.) Earth was not an element because it could be made from water, as van Helmont demonstrated to his own satisfaction. He weighed a willow seedling, put it into a tub with 200 pounds of earth, and watered it for five years. Then he removed the tree and found that its weight had increased to 169 pounds from its original five pounds. The weight of the earth, however, had remained the same. Therefore, the 164 pounds of wood that was produced must have come from the water that he had added to the pot, day by day. (This experiment had originally been suggested by the medieval polymath, Nicholas of Cusa [1401–1464], although van Helmont may have thought of it independently.) Ironically, the real source of the increase in the weight of the tree was carbon dioxide, a gas he himself discovered.

The word discovery is an odd one to apply to van Helmont's observation of the gaseous state. After all, for centuries, thousands of alchemists and chemists could not have failed to notice gases bubbling off when metals were dissolved in acids or when acids were used to dissolve limestone or any other carbonate. In fact, some three thousand years before van Helmont, the Assyrians reported that vinegar produced bubbles from lime. Nevertheless, it must be conceded that van Helmont did indeed discover the gaseous state of matter. The thousands of other chemists and alchemists who must have noticed gas bubbles had paid no attention to them.

Before van Helmont, such bubbles were nothing to get excited about. In certain reactions, the bubbles were obviously excess air. Any educated person knew that all things were composed of air, as well as fire, earth, and water. Van Helmont, however, had a new theory: All solids and liquids were made of a modified water and *not* of earth, air, fire, and water. The bubbles that he observed when metals were dissolved in acids therefore posed a real problem for him. If they were indeed air, then the Greeks were right and he was wrong. To prove that he was right, he had to study the bubbles, to show that they were not air. He smelled them, tasted them, and tried to dissolve them in water and other solvents. On the basis of taste, odor, color, solubility, and everything else he could think of, he concluded that they were not air. He was now satisfied that his theory was right and that liquids and solids were indeed made of water without any air.

Still the bubbles had to be explained. If they were not air, what were they? Some variety of air? A form of water? He finally decided that they could be liquefied by intense cold, and because he could liquefy water vapor but not air, the bubbles were simply a form of water. To distinguish between the bubbles and ordinary water, he named them *gas*, possibly obtaining the term from the Greek word *chaos*, which Paracelsus had used a century before. Van Helmont's conclusion that gas was modified water could perhaps be put under the heading of wishful thinking. Because he was never able to isolate any of his gases, he had no real experimental evidence for their liquefaction, the key to his argument. On the other hand, the British school of pneumochemists, headed by Boyle, reached exactly the opposite conclusion with pretty much the same lack of experimental evidence. They decided that gas was modified air and used the term *air* where we use the word *gas*.

In any event, van Helmont was sure that his gases were not ordinary air and that, in fact, atmospheric air was itself a mixture of gases. He believed that the atmosphere was a partial vacuum and that smoke and fire rose and dissipated into the empty spaces. Moreover, when a candle burns in a glass inverted over water, the water inside the glass rises because the burning candle removes something from the air. He produced gases in a number of different ways, for example by adding nitric acid to ammonium chloride. He tried to isolate his gases, but his receivers always burst and he never obtained any pure gases. But he could distinguish between different gases by color, odor, and taste; and he discovered what we now know as carbon dioxide, carbon monoxide, methane, nitric oxide, and sulfur dioxide. As a convinced atomist, he was sure that gases were composed of individual small atoms.

Van Helmont rejected the Three Principles theory as well as the Four Elements theory and argued cogently against them, anticipating many of the arguments of Boyle. First, he pointed out that mercury, salt, and sulfur might not be the actual principles of matter but merely artifacts produced by the action of fire during destructive distillation. Second, the materials produced by destructive distillation of substances did not recombine to form the original materials. Finally, pyrolysis, the analysis by combustion or by destructive distillation, is not reliable. For example, upon destructive distillation, some substances yield three products and others five.

Significantly, van Helmont, a transitional figure, rejected the Four Elements theory by reasoning based on Scripture but refuted the Three Principles theory by reasoning based on experimental evidence.

Van Helmont's theories had more influence in medicine than in chemistry. To explain how solid bodies or "earths" could be formed from water, he had postulated the existence of incorporeal "ferments," a word he coined. These ferments organized the water into minerals and into all living things. Applying the concept of ferments to physiology and medicine, he combined it with Paracelsus's idea of the archeus, the genius or spirit present in each organ that directs the work of the organ. To van Helmont, there were both ferments and archei, the ferment attracting the generating spirit of the archeus, which then reacts on the ferment to bring about the desired action. In the stomach, the archeus and the ferment produce acid and then, together, the ferment and the acid digest food. Although the stomach contents are acidic, other organs contain alkaline solutions. Medicines work by stimulating the archeus to send out good emanations to the other organs. There is much animism and mysticism in all this, but the ferments are akin to our enzymes, and van Helmont's knowledge of physiology was far in advance of that of his contemporaries and immediate successors. The idea of chemical digestion and the concept of the specific action of each organ are indeed correct, although not original with him.

The Next Generation of Iatrochemists

Franciscus de la Boë

In the next generation of iatrochemists, Franciscus de la Boë (1614–1672), also called Sylvius but originally plain François Dubois, subscribed to and extended van Helmont's physiology. He rejected the ferments and the archei but retained the idea that bodily functions were determined by the acid or alkaline character of the various body fluids. In a sense, this idea was a revival of Galen's humors. Evidently he was quite impressed by the violent reactions of acids with fixed alkalies (carbonates) in which there was much frothing and evolution of heat, and he made this reaction central to his physiological system. Acid and alkali were antipathetic to each other (again the age-old idea of antipathies and sympathies) and reacted with each other. The heat produced by the acid–base reaction was responsible for the life process. For example, he thought pancreatic juice is acidic and reacts with the alkaline bile, first effervescing and then forming an acid chyle. This chyle is then carried to the heart. There it is neutralized with alkali, and the neutralization produces body heat. These speculative ideas, while a bit naive to us, were recognition that the energy needed for life comes somehow from chemical reactions taking place inside the body. A century later, Black and Lavoisier would have further thoughts about this.

Otto Tachenius

Otto Tachenius, the latinized form of Tachen (c. 1620–1700), another iatrochemical physician, was an enthusiastic proponent of the acid–alkali theory of Sylvius and extended it to suggest that all substances were either acidic or alkaline. An experienced chemist, Tachenius realized that soaps were the products of an acid–alkali reaction and concluded that soap was a salt some 150 years before Michel Chevruel rediscovered the idea. In spite of his efforts, the acid–alkali physiological theory never did gain wide acceptance, at least partly because of Boyle's cogent criticism. Tachenius, who was deeply interested in industrial chemistry, lived for some time in Venice and wrote about the Venetian chemical industry, giving detailed descriptions of the manufacture of soap, ammonium chloride, mercuric chloride, rose water, and other products. These writings point up the fact that chemical industries were springing up, employing chemists not just as technicians but as supervisors, doing what we would now call "research and development." Careers not directly concerned with medicine were opening up in chemistry, and Tachenius exemplifies the many iatrochemists who were also industrial chemists.

Rudolph Glauber

Rudolph Glauber (1604–1670) was the first great chemist who was not primarily a physician. He was a brilliant self-educated industrial chemist who had advanced ideas on political economy but who still believed in alchemy. In one of his books, *Teutschlands Wohlfart* ("Germany's Welfare"), he advocated the expansion of the German chemical industry so that Germany[3] could become economically self-sufficient. This proposition reveals that by then there was not only a German chemical industry, but also that there was an incipient nationalism, a popular feeling that Germany was, or should be, a political and economic entity. Glauber was a skilled and capable practical chemist, although some of his theoretical ideas were contradictory, as might be expected in one who was essentially self-taught. In general he accepted Paracelsus's chemical philosophy, but to him the principles were not impalpable essences but actual chemicals.

Glauber earned his living by producing chemicals for sale, and he developed more efficient furnaces so that he could run his reactions and distillations at higher temperatures. He produced fuming nitric acid, sodium sulfate hydrate (Glauber's salt), and the mineral acids, and he isolated benzene and phenol from coal

[3]Germany essentially meant the German-speaking portions of the Holy Roman Empire, a loose aggregation of principalities that included modern Belgium, Czechoslovakia, Austria, Yugoslavia, Hungary, Germany, and Switzerland, as well as parts of France and Italy.

tar. He also used flame tests in qualitative analysis. Glauber, like Beguin, from whom he probably got the idea, realized about 1647 that the reaction between antimony sulfide and mercuric chloride was a double decomposition. Nevertheless, he still believed that each chemical was composed of a combination of sulfur, mercury, and salt. Somehow, he could not see the contradiction between his own ideas and observations and the classical theory he accepted. In the next generation, Boyle did see the contradiction and faced up to it.

Robert Boyle

Robert Boyle (1627–1691) bore most of the earmarks of today's chemist. He had a doctorate (an honorary M.D. from Oxford) and he was a member of a recognized scientific association, the Royal Society. He corresponded with scientists all over Europe, sending them reports of his own work and in turn passing along the information he received from them. He published in English, presented papers, and gave lectures on his work, doing all that he could to popularize chemistry. He had his own laboratory, hired assistants and secretaries, gave the equivalent of research grants, and not only did experimental work but also supervised that of others. Significantly, Boyle, the son of the Earl of Cork and a member of the highest circles of British aristocracy, called himself a chemist, and named his best-known book, published in 1661, *The Sceptical Chymist*. Chemistry was now a recognized and respected discipline.

Although he, like Glauber, still believed in the possibility of transmutation, in most other respects Boyle thought like a chemist. His attitude toward experimental work was completely modern. His techniques were quantitative, and he relied explicitly on the balance. As one of the pioneers of physical chemistry, he made quantitative measurements of the pressure and volume of air samples and found that there was a mathematical relationship between them at constant temperature. This relationship, expressed as pressure x volume = a constant, is still known as Boyle's law, the first application of mathematics to chemistry.

He did an enormous amount of work in analytical chemistry, being perhaps the founding father of that subdiscipline. He emphasized analysis by solution chemistry, rather than by the age-old pyrolysis, and developed many new methods. He used the microscope as an analytical tool, identifying crystals by their shape. He extended the use of color reagents, flame tests, and indicator dyes for acids and bases. He classified acids, on the basis of chemical reactions, as those materials that turned syrup of violets red and caused limestone to effervesce. Alkalies were materials that turned syrup of lilacs green and reacted with corrosive sublimate to give a yellow precipitate.

It is worth taking a second look at Boyle's definition of acids and bases. He classified them on the basis of chemical behavior rather than physical appearance, on the way substances react rather than on their color, hardness, or volatility. Moreover, his definition was operational. To paraphrase Boyle, if one adds a chip of limestone to a liquid and the limestone dissolves with effervescence, the liquid is an acid. Such an operational definition is totally different from the postulated, almost geometrical, definitions in use previously, like van Helmont's flat assertion that water is an element. Boyle's definition could have been produced only by someone who thought along experimental lines. A change had taken place in the way that chemists looked at chemicals.

Boyle devoted much thought and effort to the problem of chemical composition and reaction and found all of the previous theories equally wrong. Like van Helmont, he rejected Aristotle's Four Elements theory and Paracelsus's Three Principles theory, but two generations after van Helmont, he based his arguments entirely on scientific observation and reasoning and not at all on theology. Boyle argued that although wood produced flame, smoke, water, and ashes when it burned, that fact did not mean that wood was composed of fire, air, water, and earth. He pointed out that: (1) there was no proof that the smoke, flame, ash, and water existed in the wood before it was burned; they might just as well have been formed by the process of combustion; (2) there was no proof that the flame, smoke, water, and ash were simple elements; they might be even more complicated than the wood from which they were produced; and (3) when dry wood was heated in a retort, it gave off oil, gas, vinegar, water, and charcoal, but no flame; when, instead, the same sort of dry wood was heated in a fire, it gave only ashes, soot, and flame. From the retort, there were five products; from the fire, three. In neither case were there four products—the earth, air, fire, and water of the Greeks. From then on, no one could seriously offer a logical defense of the Four Elements theory, although it continued to have adherents for at least another century.

His arguments against the Three Principles theory were just as powerful, but no more effective. His first attack was against its foundation, the method of analysis by pyrolysis. The Three

Principles were the generally observed products of pyrolytic analysis: something burned, something distilled over into the receiver, and something neither burned nor distilled. Boyle now argued, like van Helmont, that pyrolysis was not a valid method of analysis and that no conclusions could be drawn from it. Glass, for example, was known to be made from alkali and sand, but no matter how strongly it was heated, it did not break down into alkali and sand, or into anything else, for that matter. In fact, if pyrolytic analysis was to be taken as the criterion, glass would have to be considered an element. As another example, fat and alkali combined to form soap but when soap was heated in a fire it did not break down into the original fat and alkali. Furthermore, when soap was heated or burned, the products were not only totally different from fat and alkali, but they could not be used to regenerate the soap. Obviously the fire did not break the soap down into its original constituents, but changed it into something else.

Boyle's arguments showed that pyrolysis was not a reliable method of analysis. In fact, combustion was not as powerful an analytical tool as was solution chemistry, using reagents such as acids and alkalies. The glass that did not break down in the flame did dissolve in alkali. After Boyle's devastating critique, any theory based on pyrolysis was open to serious question. Still, analytical chemists kept on using pyrolytic analysis at least up to the time of Lavoisier.

Another of Boyle's objections to the Three Principles theory was the fact that the three "ideal" or "philosophic" principles (mercury, sulfur, and salt) had never been found in any of the metals that they supposedly formed. This argument had been voiced hundreds of years earlier, to no avail, but the times had now changed. In the seventeenth century, chemists attached a good deal more significance to experimental evidence than did their predecessors.

Finally, Boyle declared that since two chemical substances could combine to form a compound and could then be recovered from the compound by chemical means, as in Sala's experiment, they retained their identity in the compound. Obviously such a compound is not composed of either the Three Principles or the Four Elements, but of two separate and distinct chemical substances.

These arguments could not be refuted. Instead, they were ignored. Boyle had nothing to offer in place of the older theories, and rather than do without any theory, chemists clung to something that was at least better than nothing, that is, either to the Three Principles theory or its descendant, the Phlogiston theory. In the words of the old joke about gambling, "It may be a crooked wheel but it's the only wheel in town."

Boyle also rejected the theory of van Helmont, for whom he nevertheless retained the greatest respect. He considered water to be a compound instead of an element and did not believe that it was the basic substance from which all solids and liquids were formed, even though he couldn't disprove it. He repeated van Helmont's experiment with the tree in the tub and got much the same result, but remained unconvinced. He also boiled water in a flask for days on end, adding more as it boiled away, and ultimately found a sediment on the bottom of the flask, which he cautiously suggested might indicate that the water had been changed to a solid. Yet, intuitively, he didn't believe it, suggesting instead that the sediment had come from the glass. He recommended that the experiment be repeated, and this time the glass should be weighed before and after the prolonged boiling. This is exactly what Lavoisier did a century later, and it remains a puzzle why Boyle did not do it himself. Most likely he was too busy with matters that he thought were more important than bothering to repeat a tedious procedure. This was part of his behavior pattern. While discovering what came to be called Boyle's law, he also discovered that the volume of a gas increased as the temperature increased. This is the basis of Charles's law, but he never bothered to follow that up, either.

Sylvius had suggested that body heat resulted from acid–alkali reactions. Boyle demolished that idea by pointing out that not only did some acid–alkali reactions produce no effervescence and very little heat, but also some reactions producing heat and effervescence were not acid–alkali reactions. For example, metals are not alkalies, but they do react with acid to produce large quantities of both heat and gas. Actually, going beyond an attack on Sylvius's acid–alkali theory, Boyle rejected the entire idea of affinities and antipathies as anthropomorphic and therefore unscientific. As his contemporaries pointed out, Boyle excelled at refuting theories, although he was less successful in promulgating his own.

He rejected fire as an element but not as a substance. He thought it was composed of some sort of extremely subtle atoms that had mass but could pass through glass. In this he was reasoning mostly by analogy with the behavior of light, which passed through glass and which both he and Newton believed to be composed of corpuscles that had mass. But he did come up with some experimental justification for his belief that fire consisted of atoms. When he heated quicklime in his laboratory he observed that it increased in weight and decided that the increase came from the fire, in the form of fire atoms. (Actually, quicklime absorbs carbon dioxide on standing in air if it is below the decomposition temperature of calcium carbonate. If Boyle's furnace had been hotter, his quicklime would not have gained weight.)

Boyle now decided that metals gained weight when calcined because, like his quicklime, they absorbed fire atoms. To prove his point, he devised a very important experiment. He placed a weighed sample of tin in a flask, heated the flask to drive out some of the air, sealed it, and then heated it until part of the tin turned to a chalky powder, a calx. (He had to drive some air out of the flask because otherwise, when he heated it, the increased pressure would have burst the glass.) After the flask cooled, he opened it and heard the air rush back in. When he removed and weighed the tin powder, it was heavier than the original metal. Because no chemical substance could have come in contact with the tin while it was being heated, he concluded that particles of fire had indeed passed through the glass, just as light passed through glass, and had been absorbed by the metal. An experienced, objective, experimental scientist, he repeated the experiment again and again with different metals, different flasks, different heating times, and different degrees of heating. The results were always the same. The metals got heavier when calcined, so they must have absorbed fire atoms.

To us, it is difficult to understand how such an acute observer as Boyle could have failed to see the role played by the air in his experiments. After all, he had published *On the Springiness of Air and Its Effects* and knew that air was corpuscular in nature. He knew, too, that air was necessary for both combustion and respiration, and that when something burned, part of the air was used up. From Blaise Pascal's famous experiment on barometric pressure, he knew by 1648 that air had weight, and he could easily have weighed the flasks before and after opening the seal, the way Lavoisier did when he repeated the experiment a hundred years later. Or, with his vacuum pump, he could have evacuated the flask before sealing it and so heated the metal in a vacuum. Such an experiment was exactly the sort of thing in which he was interested.

Most likely, however, Boyle never even considered that air could have been involved in calcination because it never occurred to him that calcination had anything to do with combustion. Nothing burned, there was no flame, no vapor, no soot, and no gas of any kind formed. The surface of the metal just gradually changed color, and a powder was formed. To all appearances, there was no connection between calcination and combustion. (Only two generations later, Stahl realized that calcination was a slow combustion.) In calcination, therefore, there was nothing to indicate that air was important. Moreover, Boyle, along with Newton, Hooke, and most other English scientists, accepted Bacon's theory that heat was simply the motion of particles. Fire, therefore, consisted of small, subtle particles in violent motion. Boyle already had experimental evidence to show that fire atoms had weight. Now he had designed an experiment to show that a metal contained in a flask gained weight on being calcined even when sealed away from other chemicals. His experiment had succeeded, and so his case was proved: The weight increase came from the fire.

The Problem of Combustion

Others, too, were studying the nature of combustion, and for the next hundred years, it remained a subject of primary theoretical importance. With the development of the vacuum pump, it was recognized that combustion required air. The burning substance apparently received something from the atmosphere. Yet Greek theory had viewed combustion as the loss of something, perhaps fire. Why then did the loss of fire require the presence of air?

Between 1661 and 1674, Boyle's onetime assistant Robert Hooke (1635–1703) carried out a series of studies on combustion and respiration. He burned various materials in closed vessels with water present and found that when a significant portion of the air is consumed, as shown by the decrease in its volume, combustion stops. He also observed that animals and birds kept in a closed container die when a part of the air is consumed, although a chick could live in an atmosphere in which a candle could no longer burn. Hooke finally concluded that air contained a substance like saltpeter (sodium nitrate) that was essential for both combustion and respiration.

Boyle repeated Hooke's work on combustion but, for some reason, without water being present. In his experiment both respiration and combustion ceased without any perceptible decrease in the volume of the air[4]. So Boyle rejected Hooke's theo-

[4]Boyle's results differed from Hooke's because the water present in Hooke's experiments absorbed the carbon dioxide produced by the combustion and the respiration. At that time, however, no one could possibly have guessed this.

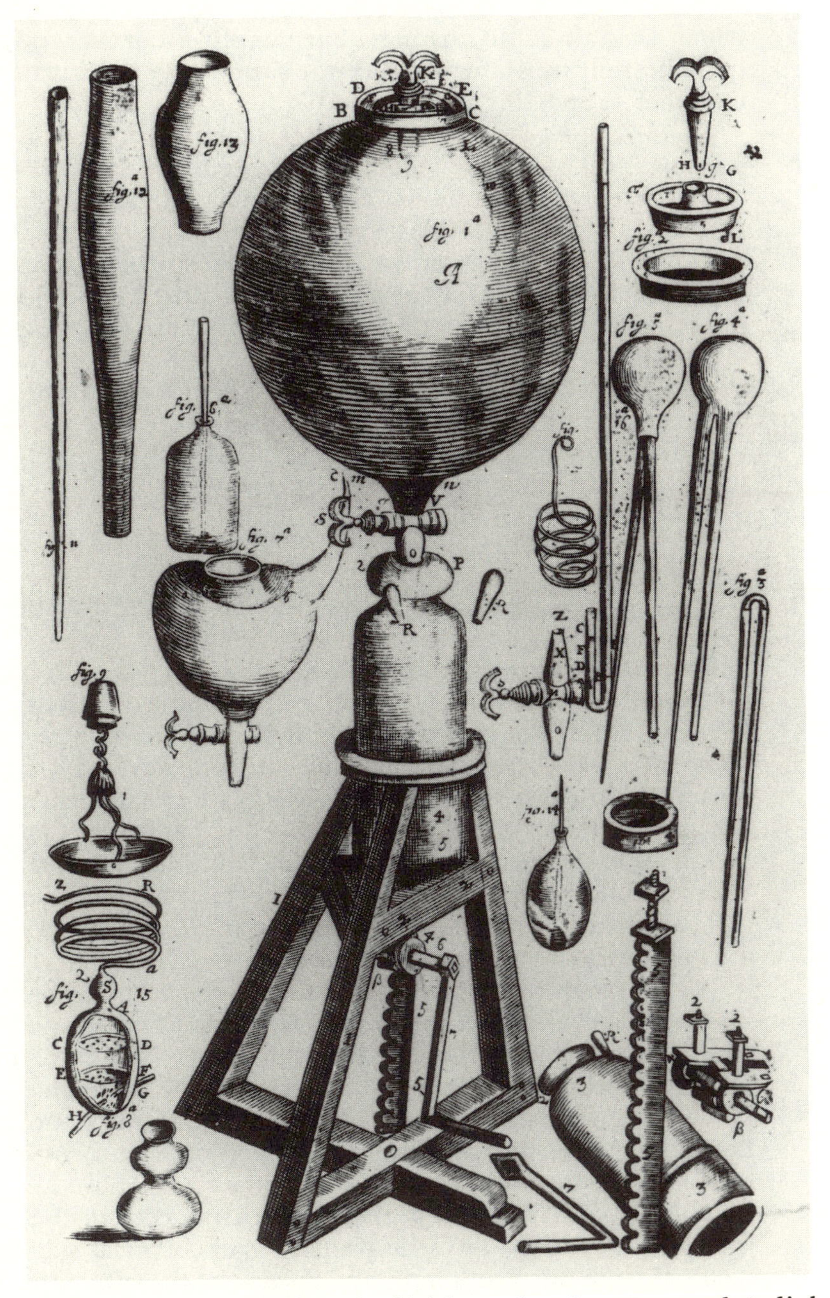

Boyle's air pump (1662), with which he produced vacuums and studied the properties of air and processes taking place under reduced pressure. (Reproduced with permission from the Bettman Archive.)

ry. Moreover, the fact that a chick could live in an atmosphere in which a candle could no longer burn implied that there might be two different substances in air, one supporting combustion and the other supporting respiration.

In 1674, John Mayow (1643–1679), whose promising career was cut short by his early death, suggested that, since a mixture of saltpeter and a combustible material will burn even in a vacuum, saltpeter is necessary for combustion and respiration. Air, therefore, must contain a "nitro-aerial spirit," consisting of fine particles too small to be seen. This is close to, but not quite the same as, the oxygen theory. It is also close to Paracelsus's idea that air contained sulfur and water and that these were necessary for life. To Mayow, the ingredient necessary for combustion was not part of the air, but something in the air. Boyle, who undoubtedly knew Mayow, rejected his theory on the reasonable ground that the known properties of saltpeter were very different from those of air.

Boyle's Concept of Elements

Boyle had been brought up with a typical seventeenth century viewpoint in which air was not really chemical in nature, although particles of chemicals might be suspended in air. Perhaps the clearest indication of his attitude toward air is his ignoring van Helmont's gases, failing to study them chemically except in a rather desultory fashion. He paid no attention to gases because he considered them as modified air, and air, of course, could be ignored in chemical reactions. Yet he knew of carbon dioxide, nitric oxide, and hydrogen, all of them gases that are obviously chemicals.

A key transition figure, he was able to demolish the beliefs of his contemporaries but unable to formulate a replacement that would be accepted by them. In Boyle's theory, the Four Elements were replaced by a "prime matter" of which all substances were made. A firm believer in a corpuscular natural philosophy, Boyle postulated that the prime matter was in the form of various corpuscles, each with its own shape and size. These he called *prima naturalis*. Chemical particles were composed of clusters or groups of particles. Physical properties of chemical substances resulted from the size, shape, and motion of the submicroscopic corpuscles. Some had points, others had holes, others were triangular. Gold, for example, was yellow because the gold corpuscles were yellow.

Boyle regarded all chemical substances as "mixts," or combinations of *prima naturalis* corpuscles. To rationalize the fact that some chemicals were made of other chemicals, he developed the concept of primary and secondary mixts, corresponding to what we call elements and compounds.

Boyle would not have accepted a chemical substance as an element. Although he was not sure what an element was, he was definite about what it was not: "I must not look upon any body as a true principle or element which is not perfectly homogeneous but is further resolvable into any number of distinct substances." In other words, if a substance can be decomposed into something simpler, it is not an element. This statement is true, but incomplete. It does not say that a substance is an element if it cannot be resolved into other substances. In our terms, it is a statement of a necessary, but not sufficient, condition. In any event, Boyle was sure that gold, mercury, and sulfur, which he could not decompose further, were primary compounds[5], not elements. Boyle's concept of an element was quite different from ours, although there has been much confusion on the subject. Part of this confusion arises from Boyle's statement:

> And now, to prevent mistakes, I must advertize you, that I now mean by Elements, as those chemists that speak plainest do by their Principles, certain primitive and simple, or perfectly unmingled bodies; which not being made by any other bodies or of one another, are the ingredients of which all those call'd perfectly mixed bodies are immediately compounded, and into which they are ultimately resolved: now whether there be any one such body to be met with in all, and each, of these, that are said to be elemental bodies, is a thing I now question [4].

Partington (*5*), following the lead of Crum Brown (*6*) and T. L. Davis (*7*), interprets the first part of this statement to mean that, to Boyle, an element was an ingredient of a chemical compound, implying that Boyle's definition was similar to ours. Marie Boas Hall (*8*), however, argues convincingly that the second part of the statement shows that Boyle's concept was quite unlike ours. In it, Boyle says that he does not believe that there are any simple bodies (elements) that are present in all substances. He contrasts his belief with the accepted view that each element is present in

[5]Boyle's concept of gold and other metals being compound substances made up of different arrangements of the *prima naturalis* made it possible for him to believe in transmutation, although not mystical alchemical transmutations by incantations and magic. If gold and lead were chemical compounds, it should be possible to change the structure of lead into that of gold. A perfectly rational idea, if one accepts Boyle's concept of *prima naturalis* clusters as being the only true atoms.

all substances. By "element" he therefore means something like the earth, air, fire, etc., of the Greeks that are present in all substances, and not our chemical elements, such as nitrogen, each of which is present in only a relatively few substances. Ironically, although he disproved the Greek theories, Boyle still retained something of the Greek conceptual framework.

He never considered that a chemical might be an element. He thought in terms of physical and chemical properties, or "qualities." These could not be explained in terms of chemical elements such as our own. For example, many substances were red and therefore had the quality of redness. Yet, there was no one chemical present in all red materials. The quality of redness, therefore, did not depend on sulfur or mercury or copper or niter or tin or any of the other chemicals. It had to be either an essence or element, which Boyle rejected, or it had to depend on a subatomic structure common to all red substances, that is, the size and shape of the *prima naturalis* particles.

Boyle's concept of *prima naturalis* generally was not accepted. It simply was not useful. It could not be tested, nor could the *prima naturalis* particles be isolated and their sizes and shapes were unknowable. Chemists continued to believe that chemical substances were composed of atoms, but continued also to accept the concept of chemical principles, in its final modifications.

Boyle was a firm believer in the quantitative approach and might have been able to change his concept of elements and qualities had he been able to study chemical reactions quantitatively. Unfortunately, this he could not do, largely because he did not know the role gases play in chemical reactions. In a reaction such as

limestone ⟶ quicklime + carbon dioxide gas

Boyle did not know that carbon dioxide was evolved because he could not see it. All he saw was quicklime being produced from limestone, and because the quicklime weighed less than the limestone, he had to believe that mass had been destroyed during the course of the reaction. On the other hand, in a calcination, such as

copper + air ⟶ copper calx

the copper calx is heavier than the original copper because it has picked up the invisible gas, oxygen, from the air. To the seventeenth century chemist who did not know of the existence of oxygen, it would seem that mass was created in the course of this reaction. Logically, therefore, although Boyle believed that *matter* was indestructible and could neither be created nor destroyed, he could not accept the idea that mass (to Boyle, *weight*) was conserved and neither created nor destroyed. Therefore he could not take the path followed by Lavoisier a century later and follow the course of reactions by measuring the masses of products and reactants. Yet without adding up the masses, there is literally no way to account for all of the reactants and products in chemical reactions or to arrive at the formulas of chemical compounds.

The seventeenth century changes in chemical thought are illustrated by the "Ode for Saint Cecilia's Day," written by John Dryden in 1687 and set to music by George Frederick Handel in 1739. In part, the ode goes

> When Nature underneath a heap of jarring atoms lay
> And could not heave her head
> The tuneful voice was heard from high
> "Arise! Ye more than dead."
> Then Cold and Hot and Moist and Dry
> In order to their stations leap
> And Music's pow'r obey.

The Four Qualities of Aristotle were still accepted, but they were now in the form of the atoms of Democritos.

References

1. Beguin, J. *Elemens de Chimie;* Edition of 1658, pp 218–219. Quoted by Hall, M. B. In *Boyle and 17th Century Chemistry;* Cambridge University Press: Cambridge, England, 1958; pp 161–162. Also, Hooykaas, P., *Chymia* **1949,** 2, 77; and Patterson, T. S. *Ann. Sci.* **1937,** 2, 243–298.
2. Sisco, A. G.; Smith, C. S. *Lazarus Ercker's Treatise on Ores and Assaying;* Chicago University Press: Chicago, 1951; p 223.
3. Partington, J. R. *History of Chemistry;* Macmillan: New York, 1961; Vol. II, p 280.
4. Hall, M. B. *Robert Boyle and Seventeenth Century Chemistry;* Cambridge University Press: Cambridge, England, 1958; p 99. Quoting from Boyle, R. *Collected Works;* 1744 edition, p 600.

5. Partington, J. R., op. cit., p 501.
6. Partington, J. R., op. cit., p 59, reference 7 from Crum Brown, A. "Development of the Idea of Chemical Composition," Inaugural Lecture, Edinburgh, 1869, 9f.
7. Davis, T. L. *Isis* **1931,** *16,* 83.
8. Hall, M. B., op. cit., pp 95–96.

XI

Phlogiston

The Last of the Old Theories

BY THE START OF THE EIGHTEENTH CENTURY, large-scale industrial chemistry was in its first growth. Unfortunately, it is almost impossible to obtain production figures because most chemical production was "captive." The chemicals produced were not for sale but were used in the course of the producer's operations, the way steel companies produce coke for use in smelting iron. We do know that a wide variety of chemicals was being produced and used, and there was a corresponding demand for industrial chemists.

New Needs, New Industries

The population of Europe was increasing. France was recuperating from the Huguenot–Catholic Wars of Religion and Germany from the devastating Thirty Years' War. The times were relatively peaceful, punctuated only by small wars and revolts. Recurrent outbreaks of plague were less severe, partly because people were better fed and had more resistance, but mostly because there was less contact with the European black rat, the host for the flea whose bite transmitted the plague. This was due partly to better housing and better sanitary conditions, but mostly to the black rat's losing out in its competition for food with the larger, and stronger, Norway brown rat. Crop failures, too, were less catastrophic, as larger and faster ships, improved transport, and a worldwide network of colonies and trading posts made it possible for food to be shipped long distances. Probably even more important was the introduction of the potato from America and the gradual spread of its cultivation (*1*). The potato has almost no waste, requires no processing, and gives a very large yield per acre. It is capable of supporting a much higher population density than the European grain crops such as barley and rye.

Living standards were on the rise as the middle class prospered, although by and large the peasantry still lived on a bare subsistence level, especially in eastern and northern Europe. Demand increased for soap, to use in the expanding textile industry and for personal hygiene because of higher standards of cleanliness. Large quantities of alkali were needed for soap and for the glass industry, which was now fabricating windowpanes, lenses, bottles (especially for wine), dishes, and drinking glasses in large quantities. Ceramic production increased in quantity and improved in quality, with chemists supervising the selection and purification of starting materials and developing new colors and glazes. Larger ships and longer voyages required improved paints and varnishes, tars, rust inhibitors, and caulking. An emerging agricultural chemical industry was beginning to produce some chemical fertilizers and chemicals for soil treatment.

Tinplate cooking utensils, dishes, flatware, buttons, and buckles were in demand. These were made by fabricating the article of iron and then dipping it into molten tin, giving it a coating or "plate" of tin. The process was a bit tricky, however, because unless the surface of the iron was carefully cleaned and all rust and scale were removed, the tin did not adhere to the iron and simply peeled off. As the best way to clean the surface was to "pickle" it with mineral acid, large quantities of acid were used in tinplate manufacture. Acids were also needed for assaying precious metals. Sulfuric acid was now being produced on a larger scale by burning sulfur in a glass bell or other large container in the presence of saltpeter (sodium nitrate) and absorbing the vapors in water. By the middle of the eighteenth century, demand for sulfuric acid was so great that in 1746 John Roebuck and Samuel Garbett in England began the "chamber" process using room-size lead containers. However, nitric and hydrochloric acids were still made in smaller quantities by heating appropriate salts with sulfuric acid.

The major use for acids was in bleaching textiles. When first spun, cotton thread is more gray than white. Cotton cloth must be bleached before being dyed or printed with patterns. The old method, used for thousands of years, had been to moisten the cloth with sour milk and then let it stand in sunlight, which generates peroxide from the water in the milk (*see* Chapter III). Although effective, this method could not possibly have been used for the enormous quantities of cotton cloth put out by the new textile machines in the last half of the eighteenth century. There simply wasn't enough sour milk, enough space for the millions of yards of cloth, or enough sunlight, especially in the sooty, smoke-darkened mill towns. Textile manufacturers tried to replace the sour milk with sulfuric acid solutions or with lime, but these could destroy the cloth and harm the textile workers and machinery. Toward the end of the century, the problem was solved by using chlorine water as the bleaching agent. This, in turn, produced a demand for chlorine and hypochlorite solutions, and a large-scale chemical industry rapidly developed. It generated great quantities of useful chemical by-products but with widespread occupational diseases and environmental damage.

The Phlogiston Theory

Despite all the chemicals and reactions now known, there was still no comprehensive theory of chemistry, no effective classification of chemicals, and no way to develop useful processes except by trial and error. Strangely, the first fairly successful attempt at a new theory was basically a modification of the older ideas that Boyle had refuted. It was proposed by Johan Becher, who for a time had worked with and for Boyle.

Johan Becher (1635–1682) during an up-and-down career had been a professor of medicine at Mainz and court physician to the Elector of Bavaria. In 1669 he proposed a Five Elements theory, based on the age-old observation that during combustion something burns, something distills, and something else neither distills nor burns. Becher's elements were air, water, and three different solids that corresponded to the three different responses to combustion. Vitreous earth (*terra prima* or *terra vitrescibile*) was glassy and fire resistant. It corresponded to the inert residue. The combustible or fatty earth (*terra secunda, terra pinguis*) was what burned. The volatile fluid, the flexible mercurial earth (*terra tertia, terra fluida*), was what distilled. Becher explicitly rejected the Four Elements and the Three Principles theories, but his three solids certainly seem very much like Paracelsus's salt, sulfur, and mercury. In fact, Becher's theory was strikingly similar to the Chinese Five Elements theory, which by this time was well known in the West. His ideas were no great improvement and attracted little attention until his student, Georg Stahl, reformulated them.

The Elements: West and East

Empedocles	Paracelsus	Becher	Chinese
Fire	Fire (sometimes)		Fire
Air	Air	Air	
Water	Water	Water	Water
Earth	Three Principles	Three Earths	Three Solids
	Salt	Inert Earth	Earth
	Sulfur	Combustible Earth	Wood
	Mercury	Volatile Earth	Metal

Georg Ernst Stahl (1660–1734), a professor of medicine at the University of Halle in Germany, was court physician to the King of Prussia. He changed the name of Becher's combustible fatty earth from *terra secunda* or *terra pinguis* to *phlogiston,* a name that Hapelius and Sennert had already used (*see* Chapter X). He then used phlogiston to explain combustion, smelting, and lime burning, three reactions of the greatest economic importance.

Combustion had always been considered to be a decomposition, the loss of something. Stahl now introduced a new concept: that what was lost was phlogiston. Phlogiston was a real substance that was transferred from one chemical to another. In smelting, an ore heated with charcoal turned into a metal because phlogiston was transferred from the charcoal to the ore. On the other hand, in calcination, when the metal was heated in air, it became a powder, or calx, because it lost phlogiston. Calcination and smelting were just reversible back-and-forth transfers of phlogiston. In effect, this was the first recognition of the fact that calcination is simply a slow combustion of metal.

Phlogiston was also the ingredient that made alkalies caustic and powerful. When limestone, soda, and potash were heated to high temperatures, they changed into quicklime, caustic soda (lye), and caustic potash (potash lye) because they picked up phlogiston from the fire. When these caustics were kept in open containers at room temperature, they became mild and lost their potency because their phlogiston leaked out into the air.

The phlogiston theory, as articulated by Stahl, was a success. Despite Becher's disclaimer, phlogiston really was Paracelsus's philosophical sulfur. As such, it appealed to conservatives brought up on the older theory. Moreover, chemists could accept it as a common-sense explanation connecting smelting, calcination, combustion, lime burning, and alkalinity, phenomena that they had not realized were related. Actually, it was very useful, and people like Joseph Priestley, Henry Cavendish, and Karl Wilhelm Scheele, who believed in phlogiston, made some first-class discoveries (*see* Chapter XII).

Flaws in the Theory

From the start, however, the phlogiston theory ran into problems. Logically, coke and charcoal had to be almost pure phlogiston because when they were burned, nothing was left but a little ash. Yet coke and charcoal certainly were not alkaline. At the time, this contradiction did not seem significant and was rationalized away, but later it caused trouble. The argument advanced was that phlogiston by itself was mild and only in combination was it caustic. Other more serious problems were posed by the weight of the phlogiston, however, and these could not be explained so easily.

When metals burned, the resulting calx was heavier than the original metal. How could the loss of phlogiston cause a gain in weight? The first response was to deny the increase in weight. Actually, until the last quarter of the eighteenth century, there was considerable controversy about this weight gain, some chemists maintaining that only lead, and perhaps tin, gained weight on calcination. The next suggestion was that phlogiston had a negative weight, so that as the metal lost phlogiston, it got heavier. This argument was clearly unsatisfactory. Priestley and others never accepted the idea of negative weights. Their explanation was that phlogiston only seemed to be lighter because its density was less than that of air. A further complication was the fact that when coke or charcoal burned, the resulting ash weighed less than the starting material. Could it be that there were two different phlogistons, one heavier and the other lighter? Eventually there were so many problems that what had originally been a simple idea became extremely cumbersome. The phlogiston theory, however, was never directly refuted nor disproved. There was literally no way to disprove the existence of an impalpable essence, that is, to prove the negative. Instead, Joseph Black showed that phlogiston was not involved in lime burning, and then Lavoisier put forward a simpler, less complicated, and more productive alternative.

There has been much discussion of whether the phlogiston theory helped or hindered the advance of chemistry. As usual in such controversies, much may be said on both sides. (Otherwise there would be no controversy.) It is clear that before the ideas of chemical structure and chemical reactions could be developed, the concept of principles like phlogiston had to be discarded. On the other hand, it is also clear that the definitive work of Lavoisier and Dalton required knowledge of the properties of gases, and this certainly was not in any way hindered by the phlogiston theory. In fact, both Black and Priestley made their discoveries while believing in phlogiston. So did Cavendish and Scheele and even, to a certain extent, Lavoisier. In any event, the phlogiston theory, although it turned out to be incorrect, provided a unifying framework that could and did suggest experiments that resulted in important discoveries.

Joseph Black

The first blow to phlogiston was struck inadvertently by the great Joseph Black (1728–1799), a physician who taught anatomy and chemistry at Glasgow University and then became professor of chemistry at the University of Edinburgh. Black was a quiet, self-effacing genius, not at all given to trumpeting his discoveries. He published only a few works other than his thesis, but his students copied down and later published his lectures. Black's attitude was in marked contrast to that of his contemporaries, at least partly because his financial and social standing did not depend on his achievements in research. He supported himself by practicing medicine, and although his professorship was unsalaried, each student paid him a fee. Other scientists were not in such a favorable situation. For them, prestigious, lucrative appointments depended on professional standing. As a result, claims of discovery and quarrels over priority became endemic in chemistry in the eighteenth century, and their frequency and ferocity have not yet abated.

Black did his first research at Edinburgh University for his doctorate and he had problems with his thesis committee. When his thesis was finally completed, his results were published in a recognized scientific journal. With these familiar hallmarks on his work, there is no doubt that Black belongs in the modern era. Among his other achievements, Black was one of the founders of the science of thermodynamics. He discovered heat capacity and latent heat, and probably helped James Watt with the steam engine. He also discovered the connection between combustion, fermentation, and respiration. He believed in phlogiston but was one of the first chemistry teachers to inform his students of Lavoisier's system of chemistry.

Between 1752 and 1754, for his M.D. thesis, young Black was searching for a solvent that would dissolve kidney stones, which cause agonizing pain. He knew that a solution of caustic alkali or quicklime would dissolve kidney stones but would also dissolve bladder tissue and kill the patient. Perhaps thinking to modify its causticity, Black started to investigate the properties of quicklime and immediately ran into a familiar problem. Two of his professors were experts on lime and had opposing views. Robert Whytt preferred lime made from oyster shells, while Charles Alston preferred lime made from limestone. Whichever variety of lime he chose, Black ran the risk of antagonizing one or the other of them. Adroitly, he switched to a different but somewhat similar material, *magnesia alba* (magnesium carbonate).

Reactions of Lime

To understand Black's work, some facts about the reactions of lime must be kept in mind. First, when limestone is heated at a high temperature, it gives off an invisible gas (carbon dioxide) and becomes a strongly caustic material, quicklime. We formulate this reaction as

$$\underset{(CaCO_3)}{\text{limestone}} \xrightarrow{\text{heat}} \underset{(CaO)}{\text{quicklime}} + \underset{(CO_2)}{\text{carbon dioxide}}$$

Quicklime is limestone minus carbon dioxide. According to the phlogiston theory, however, quicklime is limestone plus phlogiston; when limestone is heated, what happens is

$$\text{limestone} + \text{phlogiston} \xrightarrow{\text{heat}} \text{quicklime}$$

Second, when quicklime is exposed to air, it absorbs carbon dioxide and becomes limestone once again, reversing the reaction. The phlogistonians explained the regeneration of limestone by saying that the phlogiston leaked out of the quicklime back into the air.

Finally, when water is mixed with quicklime, a good deal of heat is produced, and a milky suspension of slaked lime (calcium hydroxide) is formed. Slaked lime is somewhat less caustic than quicklime, but still strongly alkaline. When slaked lime is mixed with a mild alkali, such as soda (sodium carbonate) the reaction is a double decomposition, forming limestone once again along with a strongly caustic solution of lye (sodium hydroxide):

$$\underset{Ca(OH)_2)}{\text{slaked lime}} + \underset{(Na_2CO_3)}{\text{soda}} \longrightarrow \underset{(CaCO_3)}{\text{limestone}} + \underset{(NaOH)}{\text{lye}}$$

In this reaction, carbon dioxide has been transferred from the soda to the lime. The phlogistonians, however, believed that the lime transferred phlogiston to the soda. Their concept was that of two connected reactions:

$$\text{slaked lime} \longrightarrow \text{limestone} + \text{phlogiston}$$

followed by

$$\text{phlogiston} + \text{soda} \longrightarrow \text{lye}$$

Properties of Magnesia Alba

Black began his work by studying the medical properties of magnesia and found immediately that, although it did not dissolve kidney stones, it was a very good physic. (More than two hundred years later, citrate of magnesia and milk of magnesia are still widely used laxatives.)

Now he had to learn the chemical properties of magnesia. This gave him considerably more trouble. His line of investigation and the results were roughly as follows (2).

He knew that magnesia, like limestone, reacted violently with acids, accompanied by a good deal of effervescence. Perhaps it was a form of lime. If so, it would react with acids to form salts. So he treated it with various acids and found that, in each case, the magnesia reacted with the acid the same way that limestone did. That is, it effervesced and formed a salt. But the magnesia salt was not the same as the lime salt. Magnesia was therefore *not* another form of lime.

He already knew that mild alkalies (carbonates) effervesced with acids the way magnesia did, so perhaps magnesia was a mild alkali. If so, it would react with slaked lime to produce a caustic solution, like lye. But when he mixed magnesia with slaked lime, he got a neutral or "insipid" solution. Magnesia was therefore not an alkali, either. It was an entity in its own right. But what sort of an entity?

He went back to lime for another try. Even if magnesia was not a form of lime, perhaps it could be changed into lime.

Limestone, on being heated, lost weight and formed quicklime. Maybe magnesia would do the same. Black now calcined a sample of magnesia by heating it to a temperature "sufficient to melt copper." The magnesia lost more than half its weight, seven-twelfths to be exact, and changed into a new material. But this new material was not quicklime. In fact, it was completely unlike quicklime. It was not soluble in water, and even though it must have absorbed a good deal of phlogiston from the fire, it was not alkaline.

Black found himself with a whole new problem in addition to the one with which he had started. What was this new material he had produced by heating the magnesia? What had he done by heating it so strongly?

His next step was to compare his calcined magnesia with the original magnesia. He treated it with the same acids that he had used on magnesia and found that he got exactly the same salts as before, but this time without any effervescence. In equation form:

acid + magnesia ⟶ salts + effervescence

acid + heated magnesia ⟶ same salts, no effervescence

At this stage Black must have felt that he had wandered into a labyrinth. The heated magnesia had picked up phlogiston. Why then wasn't it caustic? Why didn't the heated magnesia effervesce when acid was added? It had to be different from the original magnesia, but in that case why did it form the same salts? Finally, why had the magnesia lost more than half its weight when it was calcined? It should have lost some weight by picking up phlogiston from the fire, but seven-twelfths was a bit much.

The large weight loss provided a clue. Perhaps when the magnesia was heated, in addition to gaining phlogiston it had lost something else. He repeated the calcination and found that, on being heated, magnesia did indeed give off something—a gas (or, as he called it, "an air"). With this observation, matters began to clear up a bit. It now seemed probable to him that when acid was added to uncalcined magnesia, the resulting effervescence was simply this same gas. In one case the gas was driven off by heat, and in the other case by acid. Heated magnesia did not effervesce because it had already lost its gas.

At least, he had answered one question, but on the other hand the answer itself posed a new question: What was the gas and how much of it was there in the magnesia? When the magnesia was heated and lost so much weight, how much of the weight change was due to loss of gas, and how much to the gain of phlogiston? To answer these questions, he needed to weigh the gas, but he had no way to trap it and weigh it. The only apparatus for trapping gas was Stephen Hale's pneumatic trough (*see* Chapter XII), but the gas he was producing (carbon dioxide) dissolved in the water of the trough. Hard thinking solved his problem. Black hit upon a brilliant indirect method of determining the weight of the gas. First, he weighed the acid and the magnesia separately, mixed them together and let them react. Then, after the gas stopped coming off, he weighed the mixture. The final mixture weighed less than the sum of the reactants, and the difference was the weight of the gas.

Black must have been delighted with this observation, but no doubt startled to find that the weight loss was, within experimental error, the same seven-twelfths of the original weight of the magnesia. When magnesia was calcined, the entire weight loss was due to the loss of gas. Nothing was due to the gain of phlogiston! That explained why the heated magnesia was not caustic. It had not gained phlogiston.

By now, Black was more than halfway home. He identified the gas by running it into water and testing the solution. The gas was fixed air, the gas that van Helmont had obtained from burning wood and burning charcoal. So, *magnesia alba* contained "fixed air," but the heated, or calcined, magnesia did not. That was the difference between them.

Finally, he weighed a sample of magnesia, heated it to drive off the fixed air, and dissolved the calcined magnesia in sulfuric acid. Then he neutralized the solution with excess soda and got back the same magnesia with which he started. Both the weight and properties were the same. Obviously the fixed air that he had driven out of the magnesia by heat had been restored to the calx chemically, and because it could have come only from the soda, soda, as well as magnesia, contained fixed air.

Black's Conclusions

Black's case was now complete. He concluded that:

1. Limestone, magnesia, and the mild, or fixed, alkalies such as soda contain fixed air. (They are all what we call carbonates.)
2. When these materials are heated strongly, they do not absorb phlogiston but simply lose fixed air (carbon dioxide).

3. In solution, mild alkali contains fixed air that it can transfer to slaked lime, to quicklime, and to dissolved magnesia. Phlogiston is not involved in the transfer.
4. Limewater (a solution of calcium hydroxide) will absorb fixed air from the atmosphere and form limestone again, the limewater solution becoming opaque in the process[1].
5. The loss of fixed air is what makes alkalies caustic, although for some reason calcined magnesia is not caustic even after it has lost its fixed air.
6. All carbonates react with acids to produce carbon dioxide and salts. (This is the effervescence originally reported by the Assyrians.)
7. Calcined carbonates form the same salts with acid as do uncalcined carbonates. There is no effervescence because the carbon dioxide has already been driven off by the heat.

When Black's work was published, it was a revelation to those who read it[2]. Rarely has an author accomplished so much in one work. First, he had vindicated the idea of conservation of mass. He had demonstrated that the weight lost by the reacting materials in a chemical reaction was the weight of an invisible gas that had been evolved. Chemists now realized that in all cases where the products of a reaction were either heavier or lighter than the reactants, the hitherto inexplicable weight changes were due to absorption or emission of gases. They had to discard the old "common-sense" observation that mass was not conserved in chemical reactions. (Within twenty-five years, Lavoisier demonstrated a mass balance, for the first time.)

Second, a new class of materials had been discovered and defined by the presence of a chemical, as indicated by a chemical

[1]Almost immediately, Black used the carbonate–limewater reaction to make an outstanding physiological discovery. He went to a brewery and repeated the demonstration that limewater became milky when the gas from burning wood was bubbled through it. Then he demonstrated that the gas given off from fermentation in the brewer's vats had the same effect and finally, that same day, he showed that exhaling into limewater turned it milky. In the space of a single day, he showed that combustion, fermentation, and respiration all produce carbon dioxide, so that somehow respiration is combustion within the body. Body heat is therefore the result of combustion. (*See* Chapter X for Sylvius's acid–base theory of body heat.) It is still a mystery why Black, a practicing physician who did pioneering research on heat capacities, never followed up this discovery.

[2]However, not everybody read it. Publication was not immediate: 1756 in Britain, 1757 in Germany, but not until 1773 in France, although Antoine de Fourcroy (1755–1809) apparently knew of it in 1770. Consequently, Lavoisier and his colleagues knew Black's work only by reference to it in English and German articles, and French pneumochemistry lagged far behind that of the English.

reaction. Carbonates, as we call them today, were the class of materials that when treated with an acid gave off fixed air, as shown by effervescence. Boyle had earlier used this same reaction to define an acid. Now it was turned around to define a carbonate. It is worth noting that the reaction involved in this operational definition of carbonates was characteristic of two entire classes of materials, acids and carbonates. As such it was a prototype of the classification of reactions that would be applied to the great number of chemicals now known.

Third, the reaction scheme did not involve phlogiston. Moreover, the causticity of alkalies and quicklime had nothing to do with phlogiston. The supposed transfer of phlogiston from burning coal or wood to limestone demonstrably did not take place. This, of course, did not refute the phlogiston theory, but it dealt it a severe blow. Phlogiston had been shown to be irrelevant to one of the main reactions it was purported to explain. (However, Black kept on believing in phlogiston.)

Finally, and not least important, Black focused the attention of the scientific world upon gases. For thousands of years, gases had first been considered to be air, and then either modified air or modified water. Black had shown that there was at least one chemical substance that existed normally in the atmosphere but was neither air nor water. (Oddly, in Black's lectures to his classes at Edinburgh University, he did not treat gases as a separate class of substances, although he treated liquids as a separate class.) Soon it was realized that gases were chemicals with chemical properties, and that there might well be other gases present in the atmosphere. This had to be looked into!

The results were not long delayed. Within twenty-five years Priestley and Scheele had discovered almost a score of new gases and Lavoisier had used a gas reaction to produce the long-delayed chemical revolution.

References

1. Braudel, F. *The Structures of Everyday Life;* Harper and Row: New York,1981; Vol. I, pp 167–173.
2. Black, J. "Experiments on Magnesia Alba," a translation of his M.D. thesis, Alembic Club Reprint #1; Edinburgh, Scotland, 1898.

XII

Lavoisier and the Chemical Revolution

THE SAME QUARTER OF A CENTURY that witnessed the American Revolution and the French Revolution also saw the chemical revolution, the birth of modern chemistry. This revolution was not just the consequence of a scientific breakthrough or new factual information, although it did involve important discoveries. It was a total change in thought, in viewpoint, in terminology, and in language. The chemical revolution was the work of many, but the leading figure, the great architect of the new chemistry was, without question, Antoine Lavoisier (1743–1794).

Lavoisier's work demolished the phlogiston theory, that last descendant of the Greek–Arab theories, and provided a much more useful alternative to the old ideas. He organized the scattered chemical information into an entirely new system, our modern chemistry that is already two centuries old. Lavoisier defined chemical elements as those materials that could not be broken down into simpler substances by ordinary chemical analysis. He classified compounds as acids, bases, salts, oxides, sulfides, etc., on the basis of both their composition and their reactions. He defined chemical reactions as simply the combining and recombining of these elements or compounds. Moreover, Lavoisier changed the way chemists thought and the very words with which they thought. By giving chemicals names based on their composition, he and his associates invented the system of nomenclature that we still use in inorganic chemistry.

He invented the mass-balanced chemical equation (*1*). He differentiated organic chemistry from inorganic chemistry, and he made the first tentative suggestion of structural organic chemistry by postulating the existence of organic radicals. Following up the work of Black, he was one of the founders of the science of thermochemistry, devising, with Pierre Laplace (1749–1827), the first calorimeter to measure the heats of combustion of various substances and the heat given off as animals respire. All of this, and much more, he packed into a relatively short life, which ended at the age of fifty-one.

Lavoisier and Priestley: A Contrast

Lavoisier, of course, did not revolutionize chemistry single-handedly. Most of the key pieces in the oxygen–phlogiston puzzle, for example, were provided by other men, including his antagonist, Joseph Priestley (1733–1804), that brilliant man whose background, character, and ideas were completely opposite to those of Lavoisier. Nevertheless, the intellectual synthesis and the successful public relations campaign were Lavoisier's.

Priestley, too, was a great man, but his efforts were diffused and tied to a lost cause. He wrote some eighty now-forgotten books, pamphlets, and articles on theology, history, politics, and metaphysics. On scientific topics he wrote twelve books and about fifty papers. He invented soda water and started the study of photosynthesis with his discovery that plants absorb carbon dioxide in sunlight, but he is best remembered as the discoverer of oxygen. For that, Karl Wilhelm Scheele (1742–1786), from Sweden, deserves at least equal credit. He discovered oxygen independently a short time before Priestley, but Priestley published first (2).

In almost all respects, Lavoisier and Priestley were antitheses. Priestley did not belong to the Church of England and therefore had not been permitted to study at Cambridge or Oxford. He received no formal scientific education, but entirely on his own he acquired a thorough knowledge of chemistry and became an accomplished experimenter. He was by turns a minister, a tutor in a private academy, and, finally, literary secretary to Lord Shelburne, the British Prime Minister. Before working for Shelburne, Priestley had relied on the sales of his books on chemistry, electricity, and optics to supplement his meager income. He wrote many of the books with the commercial market in mind. Although Priestley was foolish enough to bite the hand that fed him, attacking his employer's political policies, Shelburne was magnanimous and retired him with a pension large enough for him to be able to experiment and write. Priestley was an anti-establishment figure, a radical in almost every aspect of his life but, oddly enough, a conservative in chemistry.

Unlike Priestley, Lavoisier had a solid scientific training, the best available. He studied geology, chemistry, astronomy, mathematics, and botany, all his instructors being members of the Royal Academy of France. He was a businessman, astronomer, geologist, tax collector, and chemist. He inherited money, made money, and married an heiress. He was a member of the French financial and government establishment, but in chemistry, he was a radical.

Priestley, as a typical English scientist, simply followed his curiosity without either help or hindrance from the government. Lavoisier spent much of his career in government service. Priestley was a gifted, brilliant experimentalist, but not really a theoretician. Lavoisier, although a first-rate laboratory man, was primarily a theoretician of genius. He excelled at intellectual synthesis of other men's experimental work as well as of his own. Although Priestley discovered oxygen, its significance eluded him (as it did Scheele) because of a firm commitment to phlogiston. Lavoisier, too, did not at first realize the significance of oxygen, although even before he heard of it he felt that something was wrong with the phlogiston theory. Regrettably, when Lavoisier did understand the role of oxygen in combustion, he claimed its discovery as his own. Priestley's justifiable resentment of Lavoisier probably led him to reject oxygen as the active element in combustion and calcination. He never accepted the new chemistry, and until he died, a decade after Lavoisier's execution, he retained his belief in phlogiston.

Both men suffered sad fates. Lavoisier, although sympathetic to the ideals of the French Revolution, was too closely identified with the Old Regime as a tax collector and as a government scientist. In 1794, during the Reign of Terror, he was guillotined on trumped-up charges, even though his work on improving the quality of French gunpowder was vital to the French war effort. (Two years after his death, the government was erecting statues to his memory.) Meanwhile, in Britain, Priestley had publicly sympathized with the French Revolution and so was regarded with anger and suspicion. A mob drove him from his home and burned his library. Eventually, he went into exile in America, where his friend Benjamin Franklin helped him settle, but he remained homesick for England until he died.

Hales's Pneumatic Trough

Often new theories and modifications in old theories flow from the results of inventions. So, too, with Lavoisier's chemistry. The beginning of the end for phlogiston and for the old ideas of chemical combination and reaction was the invention of the pneumatic trough by the English minister Stephen Hales (1677–1761), some time before 1727 (3). Hales believed that many materials contained air in an inelastic (nongaseous) form and that if these materials were heated, air would be evolved. To collect this air in order to prove his point, he invented the first pneumatic trough. In its original form, it was simply a tube (actually a gun barrel) leading from the area of heating to an inverted bottle full of water. When a substance was heated, any gas generated passed through the tube and bubbled up through the water into the inverted bottle. As the gas accumulated at the top of the bottle, it pushed the water out of the open end. When all the water was gone, the bottle, full of gas, was simply stoppered. With this brilliantly simple system, Hales was able to collect gases quantitatively[1].

Of course, Hales could not collect any gas such as carbon dioxide that dissolved in water. In fact, twenty-five years after Hales had published his book, *Vegetable Staticks*, Black couldn't isolate the carbon dioxide given off when he calcined magnesia (*see* Chapter XI). But the ingenious Priestley, on becoming interested in gases, modified Hales's pneumatic trough by substituting mercury for water. Now he could trap even those gases that would dissolve in water. In a short time, by heating various salts either with flame or in furnaces or with a large magnifying glass and sunlight, he produced and isolated alkaline air (ammonia), acid marine air (hydrogen chloride), dephlogisticated nitrous air (nitrous oxide), nitrous acid air (nitrogen dioxide), dephlogisticated air (oxygen), heavy inflammable air (carbon monoxide), vitriolic acid air (sulfur dioxide), and fluor acid air (silicon tetrafluoride).

Dephlogisticated Air

In 1771 Priestley obtained impure oxygen by heating nitrates but mistakenly thought it was nitrous oxide. In 1774 he got impure

[1]Hales reported that when he heated minium (lead oxide), a cubic inch of minium gave off thirty-four cubic inches of "air." He thus anticipated Lavoisier by some fifty years, although nobody paid much attention at the time (4). Hales never isolated or analyzed any of his gases, simply assuming that they all were air.

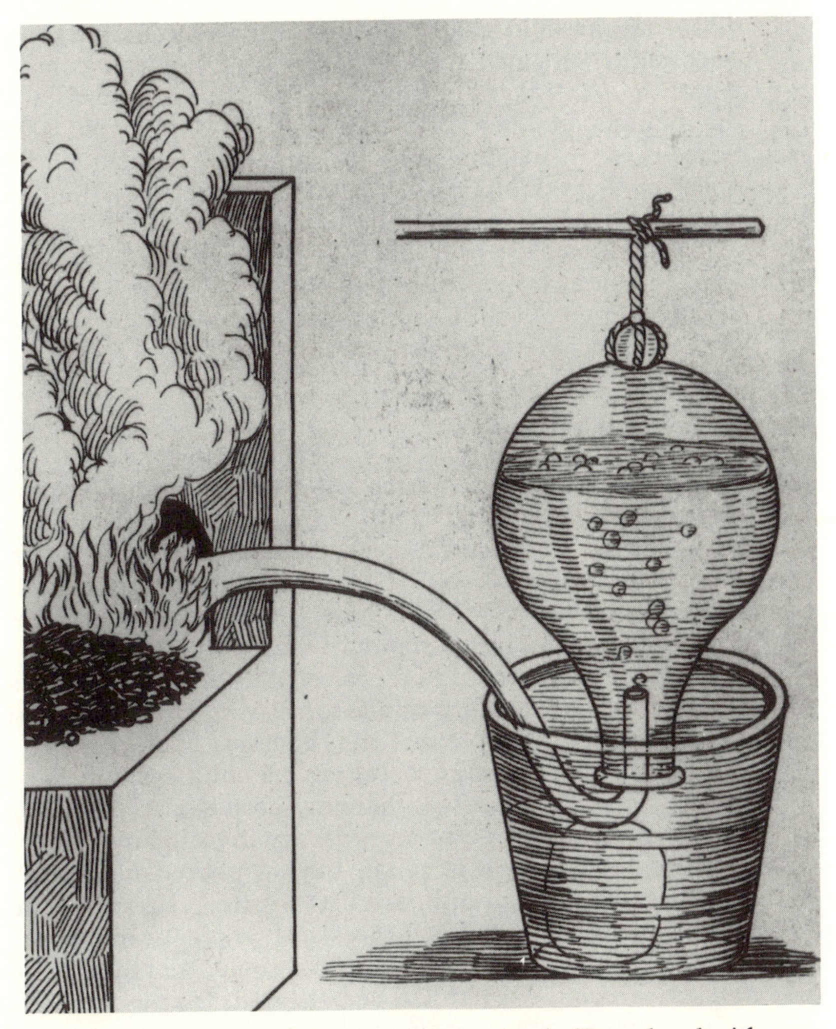

An artist's concept of Hales's pneumatic trough. (Reproduced with permission from the Bettmann Archive.)

oxygen again, this time by heating mercuric oxide, and again thought it was nitrous oxide. Finally, late in 1774, he realized it was not nitrous oxide and early in 1775 decided that he had really produced "dephlogisticated air." According to phlogiston theory, combustion stopped when either the burning substance had lost all its phlogiston or when the air was saturated and had no more room for the escaping phlogiston. (Somewhere along the line, the concept of phlogiston had changed. Originally Becher had defined it as a solid, a fatty earth. Now it was a gas that could mix or combine with air.) Priestley's new air supported combustion better than ordinary air did, so it must have had more room for the phlogiston; therefore, it had to be dephlogisticated air. Scheele, too, reasoning along the same lines, concluded that his discovery was dephlogisticated air. Neither man doubted the reality of phlogiston.

By that time, however, Lavoisier had decided something was wrong with at least part of the phlogiston theory and was investigating combustion and calcination, although without the skills and techniques of Priestley.

Lavoisier's Early Work

Lavoisier had started his scientific career with an attempt to improve the Paris water supply. In connection with that work, he measured the density of water, using distilled water as a standard for comparison. Boyle, however, had reported that there was always some solid material present in distilled water. Any solid present in his standard would invalidate his results, so Lavoisier was forced to repeat Boyle's work to see if distillation of water really did produce some sediment (5). He distilled and redistilled water repeatedly, and was finally able to show that, with careful work, there was no solid residue in the receiver. Distilled water was therefore perfectly acceptable for his standards. At that point an ordinary investigator might have stopped, with a sigh of relief, having proved his point. But Lavoisier was not an ordinary investigator. He still wanted to know what Boyle's sediment was and where it came from.

Perhaps Boyle's solid residue had come from the glass of the flask in which the water had been heated. To check this hunch, he weighed an empty flask and boiled distilled water in it for a very long time (101 days) until he finally saw a sediment. Then he filtered off and weighed the sediment, plus the residue left when he evaporated the filtered liquid, or filtrate. Then he weighed the empty flask again. As he expected, the flask had lost weight, and within a large experimental error, the weight lost by the flask was the same as the weight of the sediment plus the residue from the evaporated filtrate. (Boyle had actually suggested this experiment but never got around to doing it.) Lavoisier was now satisfied not only that his own distilled water standards were good, but also that Boyle was wrong, that water had not been and probably could not be converted into earth. His contemporaries were not so easily convinced, and as late as 1800, chemists were still doing experiments to prove van Helmont and Boyle right and Lavoisier wrong (*6*).

The practice of basing his arguments on quantitative measurements, especially as shown by the balance, was characteristic of Lavoisier. Quantitative relationships, especially the conservation of mass, were prime factors in his reasoning. He knew that he had to end up with the same weight that he started with[2]. In fact, he stated explicitly that the total mass of the reactants had to equal that of the products, and he often used a mass balance to follow the course of a chemical reaction, which he considered to be an algebraic equation. This was one of the first explicit statements of the conservation of mass, and the first expression of a balanced equation.

Combustion and Calcination

At the start of his work on calcination and combustion, Lavoisier had no quarrel with the phlogiston theory. As late as the spring of 1775, he still considered phlogiston to be real, although not necessarily involved in calcination and combustion. Still, as early as 1773 he had begun to have some doubts, and in 1774, in his *Opuscules Physiques et Chimiques* ("Physical and Chemical Tracts"), he clearly implied that he disagreed with "the disciples of M. Stahl." Working slowly and methodically, he did not offer his own theory until 1777, and not until 1783 did he present his *Reflections sur le Phlogistique* ("Reflections on Phlogiston"), a devastating critique of the problems of phlogiston.

It is not certain what first turned Lavoisier's attention toward combustion and calcination. Some authorities (Guerlac [*7*] and Siegfried [*8*]) consider that it was the work of Guyton de Morveau (1737–1816) on the weight changes during the calcination of metals. Because of contradictory reports by different investigators, there had been considerable confusion about whether all metals gained weight on calcination or some gained and some lost weight. Finally, in the summer of 1772, de Morveau had once and for all proved that all metal calxes were heavier than the corresponding metals. Lavoisier learned of de Morveau's work and found that it contradicted the impression he had received from Hales's book, namely, that all metals contain air and that when heated they evolve air. To resolve this contradiction, Lavoisier began to study combustion and calcination.

---

[2]His experience with double-entry bookkeeping as a tax collector stood him in good stead scientifically, although it later cost him his head.

Other authors take a somewhat different view. Kohler (*9*) believes that Lavoisier's experiments were undertaken not to investigate combustion and calcination but to study the acidity of certain "airs." Both Kohler and Crossland (*10*) pointed out Lavoisier's great interest in the gaseous state and in acidity. In fact, in 1777 Lavoisier named the key element in his system oxygen, which in Greek means "acid-former," rather than giving it a name that suggests combustion or calcination. By late summer of 1772, he had become acquainted with the observation of François Rozier that wine became acid on standing in air. In investigating the connection between air and acidity, he was led toward a crucial observation.

Whatever turned Lavoisier's attention toward combustion and calcination, the experiments he made in the fall of 1772 kindled doubts about phlogiston and produced the first faint glimmerings of the new theory. On October 10, 1772, he burned a sample of phosphorus in air and observed that a large amount of white smoke (phosphorus oxide) was formed. Presumably, this was phlogiston, or some form of it. He noted that these fumes were acidic and found, to his surprise, that they weighed more than the phosphorus, although phlogiston should have been very light or have had a negative weight. Nevertheless, he used phlogiston theory to explain away the contradiction. "The phosphorus is decomposed. The phlogiston leaves it. A very large amount of air is dissolved and combines with the white vapor." At that time, he obviously had not yet noticed the parallel between his observation that there is a weight gain on combustion and de Morveau's report of a weight gain on calcination. (Hales, again anticipating Lavoisier by half a century, had reported in 1727 that air was absorbed when phosphorus burned.) It seems plain that, in early October of 1772, Lavoisier still associated phlogiston with combustion.

Yet, only a few weeks later, on November 1, in a dated and sealed note deposited with the secretary of the Académie des Sciences[3], Lavoisier reported that about October 23 he had realized that calcination of metals and combustion of sulfur and phosphorus were examples of a general reaction, that all substances gain weight on combustion or calcination, and that the weight gain comes from the "fixing" of air, that is, by the combination of the substance with air. Moreover, he reported that when litharge (lead oxide) was reduced with carbon, a large quantity of "fixed air" was formed.

[3]This was a way to establish priority without letting others in on his discovery. He could continue his investigation without alerting the competition, but if anyone else announced the same result, his dated note could be unsealed to prove that he was first.

Obviously, between October 23 and November 1, Lavoisier had burned sulfur and had reduced litharge to lead, and what he saw had struck him as being very significant. Although the ideas in the note are really only conjecture, he realized the connection between calcination and combustion, which had already been revealed by Stahl (*see* Chapter XI), and he stated clearly that the weight increase on calcination is due to the gain of something from the air.

Either atmospheric air, or something present in atmospheric air, combined with metals to produce calxes and with sulfur and phosphorus to produce acids. Lavoisier did not know just what it was, but he did note that the reduction of litharge with carbon produced fixed air. There was nothing in the note concerning the idea that air contained an acidifying principle, but even if he didn't have it in mind when he burned phosphorus and sulfur, shortly thereafter the search for an acidifying principle became the focus of his work.

Fixed Air and Calcination

By February 1773, Lavoisier heard about Black's work, for he suggested that respiration and the growth of vegetation, as well as calcination and combustion, involve combining with, or the "fixing" of, something in the air. For most of the year he repeated and checked Black's results, which had at last been translated into French, and did considerable work on the ideas set forth in the sealed note of November 1, 1772. Finally, he submitted his *Opuscules* in late 1773 or early 1774, reporting on his experiments and their implications.

By now, fixed air (carbon dioxide) was in the forefront of Lavoisier's thoughts, and he had decided that upon calcination, metals combined with fixed air to form calxes, which in turn lost the fixed air on reduction. In other words, calcination was not the loss of phlogiston, but the gain of carbon dioxide.

calcination: metal + fixed air (CO_2) ⟶ calx

reduction: calx on heating with carbon ⟶ metal + fixed air

To Lavoisier, the fixed air that appeared when he reduced the calx to metal came from the calx itself, not the carbon. In phlogiston theory, to the contrary, the fixed air was supposed to come from the carbon, not the calx. Carbon was supposed to be composed of phlogiston and fixed air and when the phlogiston was transferred to the metal, fixed air was evolved. As yet, there was nothing to show which of the conflicting theories was correct. (Actually, neither is correct.) Because Lavoisier had to use carbon to reduce the calx, both calx and carbon were present at the same time. Either could have been the source of the fixed air. The only way for Lavoisier to prove his point was to reduce a calx without carbon. But at the time he did not know of any calx that could be reduced without carbon being present.

Meanwhile, it was necessary to respond to Boyle's demonstration that metals gained weight on calcination because they, or the calxes, absorbed fire atoms. (Unbeknownst to Lavoisier, J. B. Beccaria in Italy in 1759 [*11*] and M. V. Lomonosov in Russia in 1756 [*12*] had already done so.) On April 14, 1774, Lavoisier reported to the Academy that he had repeated Boyle's work and found that Boyle was wrong. He had weighed pieces of tin and lead, put them into flasks, expelled some air, sealed and weighed each flask, and then heated it until a calx had formed. Then he weighed each flask again before opening the seal. As he expected, there was no weight increase. Fire atoms either did not pass through the glass, or else had no weight. When he opened the flasks, he found that more air rushed in than had been expelled, and the weight of each flask registered an increase. Obviously some air inside the flask had been used up during calcination. Finally, he weighed each calx, and reported that the increase in the weight of the metal during calcination was equal to the weight of the extra air that reentered the flask. Thinking in terms of either fixed air or some other acid vapor, he explained the calcination reaction as a combination of the metal and the "acid part of the air."

The phlogistonians were not in the least disturbed by Lavoisier's reports. To them, the gain in weight when tin was calcined was obviously due to the loss of phlogiston. The fact that during calcination the volume of the air in the flask decreased could be easily explained. The phlogiston given off by the tin decreased the elasticity of the air in the flask, so the mixture of inelastic air and phlogiston had a smaller volume than that of the air alone.

Reports from Priestley and Bayen

In the fall of 1774, Lord Shelburne visited Paris and took his secretary, Priestley, with him. While there, Priestley had dinner with Lavoisier and a group of French scientists and told them about his exciting new discovery, an "air" produced by heating the red calx of mercury or the red calx of lead[4]. Priestley reported that he had first observed this new gas in 1772, but thought it was nitrous air because candles placed in it burned violently and completely, much better than in ordinary air. Now he was sure it was not nitrous air, nor fixed air either, because it did not dissolve in water.

We do not know if this was the first that Lavoisier heard of oxygen. About that same time, Scheele had written him a letter in which he probably mentioned oxygen. Lavoisier never said whether he got his first information about oxygen from Scheele or from Priestley. In fact, he claimed it as his own discovery. Still, no matter how he heard of it, the knowledge that such a gas existed and that it was formed by heating red lead calx or red mercury calx was exactly what he needed. But he didn't realize it. His attention was focused elsewhere.

Lavoisier was still thinking in terms of acid airs and of metals combining with "the most acid part of the air." Whatever Priestley's new gas was, it was not an acid, so Lavoisier paid no attention to it for seven months. (Priestley's new acid, "marine air," gaseous hydrogen chloride, was another matter. Almost immediately, in October 1774, he prepared and studied this new gas.) He knew from Priestley that when either the red calx of mercury or the red calx of lead is heated, it changes back to metal and gives off a gas, even if carbon is not present. But this information simply did not ring a bell. In fact, Lavoisier probably knew it already, from the work of Pierre Bayen (1725–1798). Not until March 1775 did he repeat Priestley's experiment with red calx of mercury, and even then he probably did it in connection with Bayen's reports.

A modest and capable chemist who has never received the great credit due him, Bayen reported in February 1774 that carbonates and hydroxides of mercury (our names) can be reduced to mercury by heat alone, without the presence of carbon. This was another key piece of information that Lavoisier needed but, again, it did not strike a responsive chord. Three months later, in April 1774, Bayen reported that he had heated red calx of mercury with and without carbon. With carbon present, he produced fixed air, absorbing it in water and then identifying it

---

[4]The somewhat unusual properties of mercuric oxide and red lead oxide were central to this experiment. These form readily when the metal is heated in the presence of oxygen, but if the temperature goes above five hundred degrees Celsius, they break down again to metal and oxygen. The oxides of other metals, such as iron, tin, copper, and zinc, do not decompose until much higher temperatures are reached.

qualitatively. Without carbon, he obtained a different gas, one that was not fixed air. (At the time, he did not realize the significance of this observation, probably assuming it was ordinary air.) Finally, in the fall of 1774, he reported an additional experiment showing that, on heating, basic mercuric sulfate effervesces without any carbon being present and changes into mercury. Again and again in his reports, he asserted that the phlogiston theory was wrong, because when these mercury compounds were heated, mercury was regenerated without any possible addition of phlogiston. He pointed out that, with or without carbon, gas was evolved. In one paper he even suggested that acids contain a gas, and that metals probably dissolve in acids by the transfer of this gas. "Perhaps someday one will find that metals only dissolve in acids by the aid of an elastic fluid [gas] with which they are combined during the effervescence." (*13*)

In this extraordinary series of papers, Bayen had certainly proved that, at least for mercury, reduction of a calx did not involve phlogiston at all, that therefore the metal did not contain phlogiston, and that in reduction a gas or "elastic fluid" was evolved.

Although he carefully never mentioned Bayen, Lavoisier was well aware of Bayen's work, even before he met Priestley. He actually was present when at least one of Bayen's papers was read to the Academy, and he may even have reviewed one of them. Yet for over a year he apparently ignored them. Here, too, as with Priestley's report of a new gas, Lavoisier paid no heed. His mind was still fixed elsewhere, on "acid airs."

However, in March 1775 Lavoisier began to repeat the Priestley–Bayen experiments, starting the line of investigations that overthrew phlogiston and led to the new chemistry. Even then, Lavoisier was still looking for acid airs and thinking in terms of phlogiston. As his notebooks make clear, he was trying to prove that in reduction, fixed air comes from the calx, not from carbon. He thought the gas that evolved when he heated the red calx of mercury in the absence of carbon would be fixed air (*14*). Even when the gas he got turned out to be Priestley's new air (oxygen), Lavoisier neither abandoned phlogiston nor accepted the gas as a new substance. Instead, in April 1775, he reported that this gas was only common air, somewhat purified. Furthermore, he wrote, "Atmospheric air is a combination of an acid plus phlogiston." (*14*)

Meanwhile, Priestley had continued studying his new gas. By March 1775 he had found that in all respects it was like ordinary air, only better. By May 1775, he had concluded that it was dephlogisticated air and that ordinary air was a mixture of this dephlogisticated air and phlogiston. (Even to the conservative Priestley, air was no longer an element.) In the fall of 1775 he published his results, and Lavoisier became aware of them.

Lavoisier's Combustion Theory

As was his habit, Lavoisier studied Priestley's work by repeating many of Priestley's experiments to see for himself what happened. He also devised some of his own. Little by little, he began to rethink his ideas in light of his experiments of March 1775 and his and Priestley's new observations. By 1776 he concluded that air was a mixture of two substances, one active and the other inert. (Scheele had already discovered and reported this, but the manuscript was delayed at the printer and not published until 1777.) He also proposed his own theory of acids. All acids contain "the purest part of the air" as the acidifying principle or ingredient. Finally, by 1777, he formulated his new theory, that combustion and calcination are both just the combination of substances with oxygen, the name he gave it in 1779. Fixed air is simply a combination of carbon and oxygen. Reduction of a metal calx with carbon is just the transfer of oxygen from the calx to the carbon.

One sometimes hears the comment that Lavoisier's theory of combustion was very like the phlogiston theory except that the gain of oxygen replaced the loss of phlogiston. This statement is unjust. As Lavoisier himself insisted, phlogiston is an impalpable essence, but oxygen is a real chemical that can be weighed and measured and its presence or absence determined by quantitative chemical analysis.

Because the phlogistonians believed in the principle of combustibility, they had no trouble believing in other principles and essences. So they largely accepted the idea that oxygen was the acidifying principle. However, few established chemists accepted Lavoisier's combustion theory. (One great exception was Claude Berthollet [1748–1822], who accepted the combustion theory, but after rejecting phlogiston as the principle of combustibility, found himself logically unable to accept Lavoisier's oxygen as the principle of acidity.) Most of them admitted the validity of his experimentation, but continued to believe that phlogiston was necessary for combustion and calcination. In other words, oxygen was transferred only as a result of phlogiston transfer, the loss of phlogiston being the driving force of the reaction.

Reaction to Oxygen Theory

In 1778 Pierre Macquer (1718–1784), one of the top French scientists, wrote of "a certain person who wished to meddle in higher chemistry without understanding anything of the science." Again in the same year he wrote, "M. Lavoisier has been terrifying me for some time by a great discovery . . . which was going to do no less than overthrow the theory of phlogiston . . . However, M. Lavoisier has just published this discovery of his and I can tell you that since that time I have had a great weight removed from my chest." Macquer went on to modify phlogiston theory in the light of Lavoisier's work. He proposed that phlogiston be identified with "matter of light," a compromise that met with little acceptance. In 1784 Scheele, codiscoverer of oxygen and nitrogen, wrote, "Nitric acid composed of pure air and nitrous air. Aerial acid [fixed air] composed of carbon and pure air . . . Is it credible?"

Henry Cavendish (1731–1810), as well as Priestley, died a believer in phlogiston. Their rejection of Lavoisier's oxygen theory was not due to mere stubbornness or obtuseness. These men were great scientists with fine minds. The problem lay in the fact that Lavoisier's theory showed *what* reacted and *how much* reacted but not *why* some things reacted and others did not. For example, if combustion is merely combination with oxygen, and oxygen is everywhere in the air, why doesn't everything burn? Lavoisier offered a quantitative theory with no explanations, but Cavendish and Priestley had been trained to think in qualitative terms and to demand explanations.

Composition of Water

There remained one important argument that Lavoisier had not rebutted. When a metal dissolves in acid, inflammable air (hydrogen) is evolved and a salt is formed. When the calx of that same metal is dissolved in acid, it forms the same salt, but this time without evolving any gas. The phlogistonians had a ready explanation. The gas was either phlogiston or water combined with phlogiston, which was present in the metal but not in the calx. The phlogistonians challenged Lavoisier to explain these observations on the basis of his oxygen theory. At first he could not. Then once again a phlogistonian inadvertently came to his rescue. Cavendish, who eventually abandoned chemistry rather than accept the oxygen theory, provided the clue. He exploded a mixture of common air and inflammable air and found that

Lavoisier's gas apparatus. The mercury heated in the curved retort picks up oxygen from the container and forms a calx. The calx, when heated to a higher temperature, evolves oxygen and becomes liquid mercury again. The oxygen returns to the container over the mercury surface. (Reproduced with permission from the Bettmann Archive.)

water was the product of the reaction. (Cavendish, the conservative, had proved conclusively that the Greeks were wrong again. Water was not an element. It was a chemical produced by the combination of two other chemicals.) In 1783 Lavoisier heard of this observation (which was not published until 1784) and realized it provided his answer. Acids did not produce hydrogen from calxes because the hydrogen reacts with the oxygen in the calx to form water. The water produced had not been noticed previously because the reaction between calx and acid took place in solution, with large quantities of water already present.

After this, the phlogistonians had no substantial argument to offer, although the phlogiston theory lingered until the last old adherent died. A few established chemists did accept Lavoisier's theory, even though the thought processes of a lifetime are hard to change. (Priestley himself, for a time, wavered when he found that Lavoisier was right, that hydrogen reacted with metal oxide to form water and the metal.)

The issue was decided by the new men just entering the field with open minds. Presented with the views of both sides, they saw the evidence and chose the oxygen theory. Seeing the changes in gas volume during calcination must have been very convincing. When they heated the red calx of mercury in a retort connected to a container of oxygen over mercury, they saw mercury form in the retort, and they also saw that the volume of gas in the container increased. Lavoisier said that the gas volume increased because more oxygen was formed. Priestley asked the viewer to believe that the gas volume increased because some of the phlogiston was removed. In the reverse experiment, when the mercury was heated, becoming a red calx, the volume of the gas in the receiver visibly decreased. Lavoisier said it decreased because oxygen was used up. Priestley said the gas contracted because phlogiston was evolved. The generation of student chemists first exposed to the new theory had no trouble deciding that common sense supported Lavoisier. The rationalizations of the phlogistonians were not for them. They had no previous commitment to phlogiston; some had actually used Lavoisier's books as their basic texts, as he had intended. By 1800, the phlogistonians were almost all old men, and the phlogiston theory had almost literally died out.

Lavoisier's New Chemistry

But the end of the phlogiston theory did not, of itself, invalidate all the old Greek concepts. There had to be something to replace them. If Lavoisier had rested on his laurels after proving that combustion and calcination were just oxygenation reactions, chemists might have accepted oxygenation while still retaining the Greek theory of composition and reaction. Fortunately, being intensely ambitious, he did not let matters rest. Instead, he went on to generalize the oxygenation reaction into a grand theory that included all chemistry.

We do not know what his thought processes were, but the events of the struggle against phlogiston were decisive. Priestley and Scheele had proved that air was not an element but a mixture of chemicals. Cavendish had proved that water was not an element but a compound. Lavoisier himself had shown that Becher's fatty earth did not exist. Now he discarded all the Greek elements and their descendants. He defined elements operationally, in the way that we do today, as chemicals that could neither be produced from other chemicals nor broken down into other chemicals. He then applied the mechanism of oxygenation as a pattern for all reactions. In oxygenation of mercury, 100 grams of mercury reacts with 8 grams of oxygen to form 108 grams of mercury calx. Furthermore, 108 grams of mercury calx breaks down into 100 grams of mercury and 8 of oxygen. Going in either direction, the mass of the products equals the mass of the reactants. Chemical reactions are, therefore, just the combination, decombination, and recombination of chemicals.

When he was finished, he had a system that with major modifications has lasted until the present time. Yet even at this stage, he could not simply allow his work to speak for itself. It wasn't enough to say that the emperor wore no clothes. He had the heroic job of selling his ideas. This he did by an artful arduous propaganda campaign that included writing books, planting favorable book reviews, making scientific presentations, buttonholing potential adherents to persuade them verbally, organizing letter-writing campaigns, and getting his supporters elected or appointed to positions of power and influence (*15*). It took him a decade to succeed. In retrospect, the leap from the basic facts of combustion to a complete, self-consistent theory of chemistry took not only scientific genius, but also salesmanship of a high order.

The New Nomenclature

A hundred years earlier, Boyle had failed to overthrow classical ideas because he had nothing to put in their place. Lavoisier and his associates, however, not only produced a new intellectual synthesis, a new system of chemistry, but also developed a new terminology for it. In 1787, with Guyton de Morveau, Claude Berthollet, and Antoine de Fourcroy, Lavoisier published a book titled *Methode de Nomenclature Chimique* ("Method of Chemical Nomenclature") that established the rational system that we still use in naming inorganic compounds.

Long before Lavoisier there had been considerable dissatisfaction with chemical nomenclature. The names of chemicals were complicated and usually meaningless, often left over from alchemy, terms such as powder of algaroth, pomphlix, colcothar, turbith, galena, phagadenic water, orpiment, and realgar. If the name did have some meaning it was usually misleading—butter of antimony, sugar of lead, or flowers of sulfur. As far back as the sixteenth century, Agricola had objected to the name *litharge* for a form of lead oxide on the completely reasonable grounds that the name was a Greek word meaning silver stone, and the compound contained only lead, not silver. Boyle, Newton, Lémery, Boerhaave, and many others had specific objections to inappropriate and misleading terms. The great Swedish chemist Torbern Bergman, from 1775 until his death in 1784, did considerable work on classification and nomenclature of minerals and chemicals in correspondence with Guyton de Morveau. The time was ripe for reform.

De Morveau was the prime mover in the reform of 1787. Since 1773 he had been criticizing chemical nomenclature and suggesting criteria for a systematic terminology. Lavoisier, too, influenced by the philosopher Étienne de Condillac (1714–1780), had a deep interest in the connection between words and ideas. He wrote in *Methode de Nomenclature,* and again in the preface to the *Traite Elementaire*:

> Every branch of physical science must consist of three things, the series of facts which are the objects of science, the ideas that represent these facts, and the words by which these ideas are expressed. Like three impressions of the same seal, the word ought to reproduce the idea and the idea to be a picture of the fact and as ideas are presented and communicated by means of words, it necessarily follows that we cannot improve the language of any science without at the same time improving the science itself. Neither can we, on the other hand, improve the science without improving the language or the nomenclature which belongs to it [*16*].

In 1787 de Morveau and Lavoisier and their two colleagues, Berthollet and Fourcroy, invented a new set of words for the ideas being promulgated. They gave each compound a name that indicated the elements of which it is composed. When there are two or more compounds of the same elements, they added prefixes or suffixes to the name to indicate the relative amounts of the elements. For example, the compound of sodium and phosphorus is called sodium phosphide. The suffix *ide* shows that there are two, and only two, different elements present. If oxygen is present in addition to the other two elements, there are usually two or more possible compounds, differing from each other only in their relative amounts of oxygen. Different suffixes distinguish between them. The suffix *ate* indicates the compound has more oxygen than the compound whose name ends in *ite.* The prefix *per* indicates still more oxygen, as in sodium perchlorate, which has more oxygen than sodium chlorate. The prefix *hypo* indicates even less oxygen than in the *ite* compounds; sodium hypochlorite has less oxygen than sodium chlorite[5].

[5]Sodium chlorite and hypochlorite were compounds undiscovered in Lavoisier's lifetime. The system of nomenclature was flexible enough to accommodate the new substances that would be discovered.

With this system, the chemist knew the composition of the material by the name on the bottle and could even make an educated guess as to the best way to prepare it. He could also correlate structure with properties, for example, noting that chromates were red or orange. Moreover, knowing the composition of his materials, the chemist could, for the first time, actually predict the results of chemical reactions that had never been performed. From the old names no one could have guessed that epsom salt was formed from magnesia and vitriol. Now, knowing from the labels on the bottles that magnesia was really the base magnesium carbonate, and that vitriol was sulfuric acid, the chemist could easily guess that the reaction product would be magnesium sulfate. He could now write, in words, the reaction:

$$\text{magnesium carbonate} + \text{sulfuric acid} \longrightarrow \text{magnesium sulfate}$$

The Effects of the New Nomenclature

As we think, so we speak and write, and as we speak and write, so we think. Lavoisier's rational, informative system of nomenclature revolutionized chemistry by changing the way chemists thought of chemicals. As long as chemists thought of essences and principles, even if they considered chemicals as corpuscles or particles of some kind, they would not even begin to think of their structure and composition. But when they used Lavoisier's nomenclature, they thought of compounds as combinations of atoms. Automatically, questions suggested themselves, such as, What binds the atoms together? How big are they? What do they weigh? Does the arrangement of atoms in a compound affect its properties, and, if so, how? Questions like these are still being worked on and have not yet been answered completely, but until Lavoisier had done his work they could not even have been conceived, let alone asked.

The rational, logical system of nomenclature was quickly adopted, especially by younger chemists, because it minimized memorization. Once the system was learned, chemists needed to know only the elements present to be able to name any compound and, conversely, from the name of any compound they would know its composition. It also minimized the chances of mistaken identity. The name *galena* might mean anything or nothing, but the name *lead sulfide* had only one possible meaning.

The triumph of Lavoisier's nomenclature ensured the triumph of Lavoisier's chemical system. Thinking of chemicals in the way the names connoted actually forced the adoption of Lavoisier's views. Priestley realized this and objected bitterly, but to no effect. As he complained, with reason, to get a paper published, a chemist had to use the new nomenclature. Actually, the system of nomenclature probably contributed more to the success of Lavoisier's new chemistry than the work on oxygen.

Lavoisier's final major work, *Traite Elementaire de Chimie* ("Elementary Treatise on Chemistry"), published in 1789, was a straightforward, lucid exposition of the new system. It listed some thirty chemical elements, including among them light and heat. He simply could not completely rid himself of the "principles" in which the young Lavoisier had been trained to believe. The book is a combination research report, textbook of theory, and laboratory manual. The first part is concerned with Lavoisier's special interests: gases, combustion and calcination, and acids, and discusses the results of most of his recent experiments. The second part is a classification of neutral salts based upon chemical composition. The final third of the book is a laboratory manual describing equipment and procedures, with beautifully detailed illustrations and even tables of units and conversion factors. Within seventeen years of first publication, there were twenty-three editions in six languages, and the book had carried Lavoisier's new chemistry all over Europe and America.

One final note about Lavoisier's work. The downfall of phlogiston was probably imminent and would have taken place sooner or later even without Lavoisier. His great accomplishments were the new nomenclature and the organization of the new chemistry, with its operational definition of elements, its classification of reactions and composition, and the use of the balanced equation. He himself did not make any really outstanding discoveries of new compounds or new reactions, but he did reveal new relationships, especially new quantitative relationships. Most of the discoveries that ultimately controverted phlogiston theory were the work of Priestley, Bayen, Scheele, Cavendish, and others. Lavoisier organized these discoveries into a new intellectual synthesis. He changed the way chemists thought. After 1789 the majority of chemists still worked with the same chemicals, but when they looked at a chemical reaction they did not think of essences and principles swirling around in the flask. Instead, they pictured chemicals combining and recombining with other chemicals.

References

1. Lavoisier, A. *Elements of Chemistry*; English translation by Robert Kerr (1789), facsimile edition, Dover: New York, 1965; pp 130–131, 140.
2. Partington, J. R. *History of Chemistry*; Macmillan: London, 1962; Vol. III, p 219.
3. Hales, S. *Vegetable Staticks*, 1727, reported by Partington J. R., op. cit., p 115.
4. Partington, J. R.; op. cit., p 118.
5. Guerlac, H. "Lavoisier," In *Dictionary of Scientific Biography*; Gillispie, C. C., Ed.; Scribner's: New York, 1973; Vol. VIII, pp 71–72.
6. Partington, J. R.; op. cit., pp 380–381.
7. Guerlac, H. *Lavoisier,The Crucial Year*; Cornell University Press: Ithaca, NY, 1961.
8. Siegfied, R. *Isis*, **1972**, *63*, 69, et. seq.
9. Kohler, R. *Isis* **1972**, *63*, 349–355.
10. Crossland, M. *Isis* **1973**, *64*, 306–325.
11. Partington, J. R., op. cit., p 401.
12. Partington, J. R., op. cit., p 204.
13. Bayen, P. *Opuscules Chimiques*; Paris (1797–1798), p 313f; quoted in Partington, J. R., op. cit., pp 397–398.
14. Kohler, R. *Ambix* **1975**, 22, 52.
15. Perrin, C. E. "The Triumph of the Antiphlogistonians." In *The Analytic Spirit*; Wolf, H., Ed.; Cornell University Press: Ithaca, NY, pp 41–63.
16. Lavoisier, A., op. cit., pp xiv–xv.

XIII

Atomic Weights and Molecular Formulas

THE WORK OF LAVOISIER pointed to the new directions that research would take. For the natural philosopher, physical and chemical properties would be explained on the basis of atomic and molecular structure and composition instead of principles and qualities. The questions would now be the following: What makes an atom of one element different from the atoms of other elements? Why does a particular element combine with some elements and not with others? What is there about atoms that causes a substance to be hard or green or volatile? What are the atoms present in the various compounds? How are the atoms arranged, if, in fact, they are arranged?

For the practical chemist, or "engineer," there were other problems. Lavoisier had described the first balanced chemical equations. He had demonstrated that in any reaction the mass of the reactants had to equal that of the product. Still he could not make quantitative predictions. He could not calculate the number of pounds of salt needed to make a hundred pounds of soda. He had to work it out by trial and error. All of the thousands and thousands of numerical relationships needed by production chemists still had to be determined experimentally.

In hindsight, it is obvious that the practicing chemists needed first to determine atomic and molecular weights and molecular formulas. Once these were known, they could balance the equations that Lavoisier had established and could then calculate relative weights of reactants and products. No longer would they have to work out trial-and-error recipes for all of the innumerable chemical processes. Moreover, theoretical chemists, knowing the atomic and molecular weights and formulas of chemicals, could correlate their formulas with their properties. And from the composition of reactants and products they could work out just what had happened in the course of a reaction.

But if it is clear to us that atomic and molecular weights and formulas were the first item on the agenda, unfortunately, it wasn't as clear to chemists at that time. They were too busy hacking their way through the underbrush to be able to see the forest.

Chemists were struggling to readjust their thinking. They were redefining their objectives and thinking through the implications and consequences of all the new ideas. Even for the few who realized what had to be done, the tasks of finding both the atomic weights and the molecular formulas were insuperable. Whether or not they realized it, the problem *was* insuperable, a mathematical impossibility. It meant solving an equation with two unknowns. If they knew the molecular formulas, then with the help of good analytical data, they could easily have calculated the atomic weights. Or if they knew the atomic weights, they could easily have worked out molecular formulas. But they knew neither. In fact, they were not even certain that there were any molecular formulas. Most chemists felt intuitively that if a compound atom (molecule) did exist, it had a fixed and definite composition, but there was no proof, and, of course, unless there was a definite composition, there was no formula.

The Proust–Berthollet Controversy

The Proust–Berthollet controversy illustrates the problem chemists faced. Joseph Louis Proust (1754–1826) decided early in his career that compounds had fixed compositions and set out to prove it. While working in Spain as a metallurgist and analytical chemist, he noticed that there are two different iron sulfates, one red and the other green. Their properties contrast sharply, and their percentage compositions are quite different. Investigating further, he found no iron sulfates other than the red and the green ones. From time to time he did find sulfates of iron that seemed to have a composition intermediate between those two, but, on checking, he was always able to separate them into a mixture of the red and the green sulfates. He noticed that lead, copper, antimony, nickel, and cobalt also form two, and only two, sets of sulfates with different properties and compositions. Each of these metals also had two, and only two, different sets of sulfides. (During this investigation he originated the use of hydrogen sulfide to precipitate metal sulfides, a technique that for almost two centuries has permeated qualitative analysis laboratories with the familiar odor of rotten eggs.)

In 1797 Proust announced what we now call the Law of Definite Proportions: Each chemical compound has a fixed and invariable composition by weight and therefore a fixed and invariable composition, that is, a formula. His results should have settled any doubts about the reality of chemical composition. But they didn't. Instead, in 1799, no less an authority than Claude Louis Berthollet, Lavoisier's collaborator and coauthor, announced that a compound need not have a fixed composition and that, most likely, each substance has an approximate formula that can vary within wide limits. That precipitated an argument that lasted almost ten years, until Dalton's Law of Multiple Proportions settled the matter.

Berthollet was a great man who made notable contributions to chemistry, chief of which are his collaboration with Lavoisier on nomenclature, his studies of reversible reactions, and his seminal work on the Mass Action principle. His doubts about chemical composition stemmed from his work on chemical reversibility and equilibrium.

From 1798 to 1799, Berthollet was in Egypt as a scientific adviser with Bonaparte's ill-fated expedition. There he observed a lake of salt water with a limestone bottom and a crust of sodium carbonate crystallizing at the water's edge. To any laboratory chemist, this was astonishing. The sodium carbonate was clearly being precipitated by the reaction of dissolved limestone with the sodium chloride in the water. Yet chemists had long known that limestone is precipitated by sodium carbonate from solutions of lime salts. What Berthollet saw was the reverse of the usual reaction. He now began to wonder how many other reactions like these could be reversed and how such reversibility might be explained.

When Bonaparte and his retinue evaded the British naval blockade, Berthollet escaped back to France with them. There he made a thorough study of reversible reactions, concentrating on acid–base reactions that involved mixtures of alkalies and alkaline earth bases. He made up reaction mixtures in various proportions, let them react under all sorts of conditions, and then analyzed the products. Finally he concluded, correctly, that changes in concentrations and conditions not only change the relative amounts of reactants and products in the equilibrium mixtures, but also can even reverse the direction of a reaction, making it go backward, as in the case of the salt–limestone reaction. (This was the first statement of what eventually became the Mass Action principle.)

Up to this point, he had been right, but then his indifferent analytical technique led him astray. Every time he ran his reactions, he analyzed the products, and each time he found that the composition of any particular product was slightly different. Actually, his products were not pure. His reaction mixtures contained various impurities as well as some solid solutions. But he decided that his materials were indeed pure and that therefore the composition of his products had to be variable. Hence there were no constant proportions.

Berthollet's preconceptions had led him to see only what he expected to find. From the beginning, he probably was looking for variable proportions. Even before he started studying reversibility, he reasoned that the composition of a product should depend on the attractions, or affinities, between its components. The greater the affinity between two different elements, the more atoms there should be in the compound. After all, by analogy, the stronger you are, the more you can hold. Unfortunately, this is not true in chemistry. Oxygen atoms attract hydrogen atoms more strongly than carbon atoms do, but each carbon atom combines weakly with four hydrogen atoms, and each oxygen atom combines strongly with only two hydrogen atoms. Bond strength has no connection with valence, but in 1799 there was no way for Berthollet to know it.

In any event, Berthollet and Proust argued for years without either making the slightest dent in the ideas of the other. Finally, John Dalton's work in the early 1800s settled matters conclusively. If Dalton was right about multiple proportions, Proust had to be right about definite proportions.

Mathematical Relationships

The first tentative step toward working out atomic weights and formulas was taken by Jeremias Benjamin Richter (1762–1807), a chemist at the Royal Porcelain Factory in Berlin. Evidently industrial chemistry was now so specialized that there were porcelain chemists[1]. Richter had been a student of the philosopher Immanuel Kant and believed that all sciences were branches of mathematics. He looked for mathematical relationships between the weights of materials in chemical reactions, and he found them. For example, to paraphrase his work, to neutralize 1,000 grams of caustic potash (potassium hydroxide), it took either 873 grams of sulfuric acid, or 650 grams of muriatic (hydrochloric) acid, or 1,123 grams of nitric acid. If any other base was used instead of potash, the acid reacted in the same proportion: The amount of base that neutralized 650 grams of muriatic acid would also neutralize 873 grams of sulfuric or 1,123 grams of nitric acid. In other words, 1.00 gram of sulfuric acid did the work of 0.745 gram of muriatic acid or 1.29 grams of nitric acid. The ratio of the weight of muriatic acid to that of sulfuric acid was 0.745 to 1.00, and the ratio of the weight of nitric acid to that of sulfuric acid was 1.29 to 1.00. Every pair of acids reacted in such a fixed weight ratio, and every pair of bases used in neutralizing acids was also in a fixed ratio by weight. In the decade after Lavoisier's death, Richter published three books on such mathematical relationships between reacting materials, and he coined the word *stoichiometry*, which we still use for these relationships.

Ernst Gottfried Fischer (1754–1831) reworked Richter's tables and published them in 1802 as tables of combining weights (equivalents), based on sulfuric acid as a standard. The work was difficult to read and understand, and it made no great stir. Its implications were unclear. There has been much discussion as to whether Richter anticipated John Dalton or influenced him, but his work does not really seem to have had much influence on Dalton, or on many others beside Fischer.

Dalton and Atomic Repulsions

John Dalton (1766–1844) was the son of a poor weaver. He left school to go to work at age eleven, but was so bright that at age twelve he was hired as village schoolteacher, with some pupils older and larger than he. As a Quaker, he was not eligible for

[1]Thackray (*1*) speaks of the "professionalization of science" resulting from the Industrial Revolution. He refers primarily to Britain where William Wollaston (1766–1828), Sir Humphry Davy (1778–1829), and others could earn a good livelihood as practitioners of science in contrast to earlier eras when scientists were usually dilettantes. On the Continent, too, toward the end of the eighteenth century Proust, Berthollet, Scheele, Richter, and Bergman were all professional scientists.

scholarships to the British public schools and universities, but he educated himself in science. His first discovery was a form of color blindness, now known as daltonism.

Dalton was not a chemist but a meteorologist who got into chemistry through his interest in the composition of the atmosphere. He was troubled by the fact that oxygen and nitrogen do not form layers in air, with oxygen, the heavier gas, on the bottom[2]. He also wondered where water vapor went when it evaporated, where smoke went when it dispersed, and where water came from when it rained.

In working out answers to these questions, Dalton took as his starting point Newton's concept of gases as being motionless atoms that repelled each other. Boyle had reported that the pressure and volume of a gas increase on heating, and Newton explained this as the result of interatomic repulsions. Dalton, in turn, explained the repulsions on the basis of Lavoisier's "caloric fluid." Lavoisier had postulated that heat is a weightless fluid that he called *caloric*. Each atom has a layer of caloric around it. When a substance is heated, the caloric layer around each atom gets larger, increasing the volume of the atom and therefore decreasing the density of the substance.

Dalton suggested that the caloric around each gas atom repelled the caloric around all the other gas atoms, and thereby produced pressure. The higher the temperature, the more caloric around each gas atom and therefore the greater the volume and pressure of the gas.

The trouble with this picture, as Dalton soon realized, was that if all the atoms of gas repelled all of the other gas atoms, there would be spaces between atoms, and in a mixture, such as air, the heavier atoms would fall down between the lighter atoms. The air would, therefore, become stratified with layers of oxygen and layers of nitrogen. This does not happen, so the simple idea that gas atoms repel each other must be wrong. Rethinking his position, Dalton now hit on the happy thought that atoms of oxygen repel only atoms of oxygen, not atoms of nitrogen. The nitrogen atoms, in turn, repel only nitrogen atoms. Like repels like. Such selective repulsion would explain why air was homogeneous and not stratified.

Considering the consequences of such selective interatomic repulsions, Dalton predicted logically that in a mixture each gas

[2]Priestley had observed that mixtures of light and heavy gases do not separate on standing (2). Dalton, in early 1803, showed that even if hydrogen and carbon dioxide were placed in contact, with the lighter gas on top of the heavier, the two gases mixed uniformly.

would exert its own pressure, as if the other gases were not there at all. Dalton's Law of Partial Pressures was soon verified experimentally, and turns out to be a valid relationship based on invalid assumptions. He was right for the wrong reason. The gas molecules do not repel each other; they are simply in constant motion, moving through the empty spaces between molecules.

The First Molecular Formulas

As a result of discussions with William Henry (1774–1836), discoverer of Henry's Law of Gas Solubilities, Dalton turned his attention to the relative solubilities of gases in water. Thinking perhaps that gases whose particles were heavier or more complex would be more soluble, he looked to chemistry for evidence (3). Unfortunately, the information he wanted simply did not exist. Nobody knew which gases had heavier particles. We do not know if anyone had even thought about that question. And had any chemists tried to work out atomic or molecular weights, they would have been stopped by the problem of one equation with two unknowns, the same problem that now confronted Dalton. If he had the formula, from analytical data, he could have obtained the atomic and molecular weights. If he had the atomic weights, he could have found empirical formulas. But he had neither formulas nor weights, and so he had no mathematical or chemical way of finding either.

Dalton, however, was not a chemist. He looked at this problem from the viewpoint of a physicist who had already successfully predicted a law based on repulsions between atoms of the same element. He "knew" similar atoms repelled each other. To him, the answer was almost obvious. Because of selective interatomic repulsions, only a few simple combinations of atoms were possible. From these few possibilities it should be easy to pick out the right ones and use them to work out atomic weights. From these atomic weights, he could then get more formulas.

There have been several conjectures as to how and when Dalton got his original idea. He gave different versions on different occasions. Most likely the idea occurred to him while he was trying to picture, or even to make, three-dimensional models of possible combinations of atoms. He suddenly realized that, because of repulsions, only a few possibilities existed. In any combination of A and B, if atoms of A were in contact with other A atoms (or B with other B atoms), they would repel each other and break up the combination. Therefore, only a few atoms of A could be placed around an atom of B. An AB and even an ABA combination could exist. But if there were more than, perhaps, four A atoms around one B atom, the repulsions between A atoms would tear the molecule apart. Combinations such as A_5B could not exist. The only possible combinations of atoms were as follows:

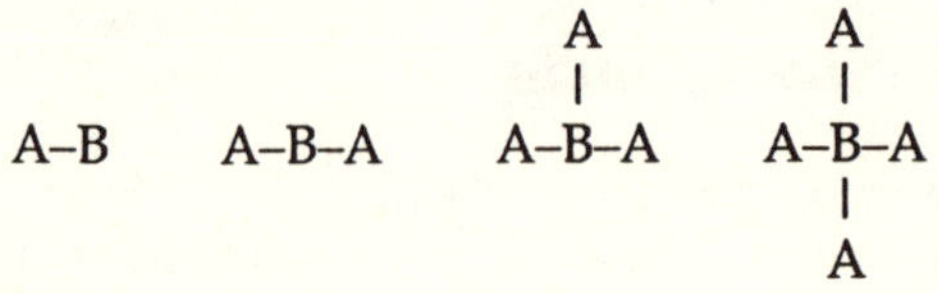

```
                             A          A
                             |          |
A–B        A–B–A         A–B–A      A–B–A
                                        |
                                        A
```

With this picture in mind, Dalton could narrow the number of possible atomic combinations down to three or four. From these he picked what he considered to be correct formulas, making his selection in the belief that, in general, nature would follow the rule of "greatest simplicity." He stated (4), "When only one combination of two bodies can be obtained it must be presumed to be a binary one, unless some cause appear to the contrary." Water, for example, would have the formula HO, with one atom of hydrogen and one of oxygen.

In most cases he was able to arrive at a plausible formula. But he soon ran into a problem. Sometimes two elements form two or more compounds, such as the two iron sulfides and the two copper oxides. How could he select the correct formulas for these? Which of the two copper oxides would be the simplest combination, and which would not be—and how could he tell? The answer was obvious to Dalton, although nobody else had ever thought of it.

Multiple Proportions

If A and B formed two compounds, such as AB and A_2B, the molecule with 2A and 1B would have twice as much A as the other. Therefore, in one compound, the weight of A per gram of B would be twice as much as in the other. If a molecule had the formula A_3B, it should have three times as much A per gram of B as would the molecule AB. And so forth. (This relationship is known as the Law of Multiple Proportions, although Dalton did not at first call it that.) Generalizing, if two elements formed a series of compounds, the weights of A per gram of B for each compound in the series should form a simple proportion of whole numbers, such as 1 to 2 to 3 to 4.

Having made up his mind beforehand, and knowing what to look for, Dalton now found evidence to support his hypothesis. He knew of three different gases formed from nitrogen and oxygen. For each of these gases, he determined the weight of oxygen per gram of nitrogen. To his great satisfaction, he found, within experimental error, the ratio 1 to 2 to 4. Even though for some reason the number 3 did not appear, he had actually found the small whole-number ratios that his hypothesis had predicted. For further confirmation, he made analytical measurements on the two gaseous oxides of carbon and found that the weights of oxygen per gram of carbon were in the proportion 1 to 2.

Dalton was now convinced he was right. Atoms did indeed combine in small whole-number proportions. Then he turned his argument around. Because he knew which oxide of carbon had the least oxygen, he could assign to it the formula CO. The other oxide had twice as much oxygen, so its formula was CO_2. (They could also have been C_2O and CO.) The nitrogen oxides had ratios of 1 to 2 to 4, so that their formulas were NO, NO_2, and NO_4 or, more likely, N_2O, NO, and NO_2, because the probability of fitting four oxygen atoms around one nitrogen atom was small. From the formulas, he now could obtain the relative atomic weights. For example, in any sample of CO, the measured ratio of the weight of carbon to that of oxygen was the same as the ratio of the weight of one atom of carbon to that of one atom of oxygen.

Dalton set to work with other gases. By 1803 he had published a table of weights of ultimate particles, with hydrogen's weight as 1, because he found that hydrogen was the lightest element. Now he could show that heavier gas particles indeed were more soluble in water than lighter ones.

At first Dalton's ideas did not make much of a stir, but they convinced Thomas Thomson (1773–1852). Other British scientists, except for William Henry, were skeptical. Thomson had written a widely used textbook[3] and in 1807, in the third edition, he published the Dalton theory, giving what he called a "table of relative density of the atoms," along with an explanation of how densities were determined. He then extended Dalton's theory, which had been applied only to gases, and used it to calculate the molecular weights of acids, bases, and salts. Finally, he reported that the Law of Multiple Proportions applied to series of solids as well as to gases, citing experimental evidence.

---

[3]Textbooks were becoming increasingly important. More and more chemists were learning their basic theory from textbooks, either formally at universities or with private tutors or else informally at home.

ELEMENTS

	W.t		W.t
Hydrogen.	1	Strontian	46
Azote	5	Barytes	68
Carbon	54	Iron	50
Oxygen	7	Zinc	56
Phosphorus	9	Copper	56
Sulphur	13	Lead	90
Magnesia	20	Silver	190
Lime	24	Gold	190
Soda	28	Platina	190
Potash	42	Mercury	167

Dalton's 1803 table of weights of ultimate particles. Each element is accompanied by its symbol and weight. Note that nitrogen is only 5 and oxygen 7. Dalton considered the alkali and alkaline earth carbonates to be elements. (Reproduced with permission from the Bettmann Archive.)

Today the Law of Multiple Proportions is so unimportant that most college chemistry texts do not even mention it. But at that time it was sensational. Multiple proportions startled the scientific community and brought the Dalton theory to its attention. Not only had a law of nature been predicted correctly, but it was a quantitative prediction. Furthermore, the law applied not only to the work of Dalton and Thomson. Anyone could use the results of previously published chemical analyses, by Proust, for example, and find additional cases. Jöns Jakob Berzelius (1779–1848) in Sweden learned of Dalton's theory from Thomson's book and immediately set to work, spending some ten years doing an enormous amount of analytical work on a great number of compounds for which he reported multiple proportional relationships. Berzelius's reports really established Dalton's theory. As a result of Thomson's book and Berzelius's work, the Law of Multiple Proportions became a fact of chemistry. So too did the Law of Definite Proportions, because unless there were definite proportions, there could be no multiple proportional relationships. This ended the Proust–Berthollet controversy in favor of Proust. Not until many years later was it discovered that some solids do have variable compositions. They are appropriately called berthollides.

Dalton's great claim to fame is his version of the Atomic theory. This, as we know it, can be stated in three simple propositions[4].

1. Atoms are indivisible and are neither created nor destroyed in chemical reactions. (This idea was at least two thousand years old.)
2. Each atom of an element is exactly like every other atom of the same element and different from atoms of all other elements.
3. When atoms combine with each other, they do so in the ratio of small whole numbers

The second and third statements were the new and important parts of the Dalton theory. Eventually it turned out that neither was quite correct, yet they were a most useful approximation.

To sum up Dalton's achievements, he did not initiate the Atomic theory. He started from generally accepted principles

[4]Dalton never published these propositions in this form. In September 1803 he wrote in his notebook eleven statements that taken together include the entire Atomic theory (5). In his papers, he restated one or more of them, from time to time, but the accepted version of the Dalton theory emerged only little by little from the discussion of his work by others.

and from them drew conclusions about atoms and molecules that could be verified experimentally. He then used these ideas to obtain some empirical atomic and molecular formulas. With these formulas, and his table of atomic and molecular weights, chemists could for the first time write balanced chemical equations in terms of the numbers of atoms reacting and the weights of reactant and product. The atomic and molecular weights he obtained were not notably accurate, nor were his formulas, but from them more accurate weights and formulas could be worked out, and ultimately the correct values. From these correct formulas, structures could be determined.

The time frame of Dalton's work is worth noting. It took two thousand years to advance from the Four Qualities of Aristotle to Lavoisier's chemical revolution. With the new viewpoint established, it took only nine years from the time of Lavoisier's death for Dalton to come up with his first table of weights.

Scientific Reaction to the Dalton Theory

It would be an overstatement to say that Dalton's work immediately revolutionized chemistry. In 1807, when Thomson first called his work to the attention of the scientific world, and in 1808 when Dalton published his own book, *New System of Chemical Philosophy*, not many scientists accepted the Dalton Atomic theory. Physicists never did like it. It was too primitive, too simplistic. It left unanswered all the questions of primary importance to physicists, such as the internal structure of the atom, the nature of binding between atoms, and the reasons that reactions occur. Physicists granted the fact that in chemical reactions substances combined in certain fixed quantities that could be called "chemical equivalents" of atoms. But this did not necessarily mean that matter existed in the form of chemical atoms. The chemical equivalents might be formed from matter during reactions by the forces involved in the reactions. Or, if atoms really existed, the physical atoms might not be the same as the chemical atoms. Furthermore, any atoms that existed certainly could not be Dalton's hard, impenetrable atoms (*see* Chapter XV). Even in the period from 1840 to 1860, when Herepath, Joule, Clausius, Waterston, and Maxwell (all physicists) calculated the dimensions, velocity, and rate of collision of gas particles (molecules), they still did not accept the reality of chemical atoms.

Some chemists, too, were skeptical. Kekulé and Frankland did not believe in the physical existence of chemical atoms, nor did Gerhardt and Kolbe. As late as the 1890s, the great organic chemist August Hofmann still rejected the chemical atom as nothing more than a convenient fiction. So did Wilhelm Ostwald, the father of physical chemistry. Still, like it or not, the concept of chemical atoms was indispensable to the practicing chemist.

But how to determine the formulas of compounds? Different investigators got different atomic weights. Which ones were correct? Was water HO, H_2O, HO_2, or H_2O_2? No one knew. Depending on the experimental method he used, an investigator might get several different atomic weights for the same element.

A quick solution to most of the chemists' problems was offered quite early by the work of Joseph Gay-Lussac (1778–1850). Unfortunately, it contradicted accepted theory, and it took half a century to straighten out the resulting chaos and confusion. In 1808, he reported that when gases react, the volumes produced and consumed are in the ratio of small whole numbers, very like the small whole numbers of atoms in Dalton's Atomic theory and in the Law of Multiple Proportions (*6*). For example, when hydrogen and nitrogen react, three liters of hydrogen reacts with one liter of nitrogen. For many other reactions, the combining volumes are also in the same sort of small whole-number relationships. The obvious implication was that the ratio of gas volumes is the same as the ratio of atoms involved, which means that equal volumes of gases contain equal numbers of atoms. Gay-Lussac called his gas particles molecules, not atoms, but he used his results to calculate formulas of muriatic acid (HCl) and the oxides of nitrogen. Applying this idea to other gas reactions, the combining volumes of gases could therefore be used to obtain large numbers of correct formulas. From these formulas, atomic weights could be calculated.

Dalton refused to accept the validity of this calculation in any manner or form. He was adamant. The most he would do, much later, grudgingly, was to say that there was something "wonderful" in the frequency with which the combining volumes were in the ratios of small whole numbers. To do him justice, he had some good reasons. To paraphrase his argument, in the reaction between hydrogen and chlorine, one liter of each gas produces two liters of hydrogen chloride, also a gas. By Dalton's definition, the particles of hydrogen and chlorine gas were atoms, although the hydrogen chloride particles were compound atoms. If equal volumes of gas do contain equal numbers of particles, there are twice as many compound atoms of hydrogen chloride at the end of the reaction as of hydrogen at the beginning. Therefore, each atom of hydrogen (and also each atom of chlorine) would have to split in two during the reaction. By the very definition of the word atom, this is impossible.

Dalton had another argument that actually predated Gay-Lussac's work. In 1803 Dalton had noted that a volume of oxygen is heavier than a volume of water vapor (4). So if the number of compound atoms of gas in a volume of water vapor is the same as in a volume of oxygen, an atom of oxygen is heavier than a compound atom of water vapor, even though the water particle contains hydrogen as well as oxygen. In other words, adding a hydrogen atom to an oxygen atom makes the pair of atoms lighter than the original oxygen atom. This is obviously absurd. Equal volumes of different gases could not possibly contain equal numbers of atoms.

Avogadro's Hypothesis

Both of these valid arguments were circumvented simply, but to no immediate avail, by Amedeo Avogadro (1776–1856). In 1811 he suggested that gases were composed not of atoms but of particles containing several atoms. These he called molecules (a term already used by Gay-Lussac and others [7]). Dalton's objections to Gay-Lussac's work could now be answered by simply assuming that hydrogen and chlorine molecules are diatomic, that is, they each contain two atoms. Therefore, a volume of H_2 contains the same number of hydrogen atoms as two volumes of HCl. The hydrogen atom does not split, but the hydrogen molecule does split into two atoms. An atom of chlorine is much heavier than one of hydrogen, so the molecule of H–Cl was of course much lighter than the molecule of Cl–Cl. Also a molecule of O–O would be heavier than a molecule of H–O.

Avogadro's arguments were completely correct. Eventually, after the first Scientific International Congress at Karlsruhe, Germany, in 1860, they were accepted by chemists. To us, it may seem strange that it took the scientific community forty-nine years to accept such a logical viewpoint, but there were valid reasons for the delay.

Even if Avogadro was right and gas molecules consisted of several atoms, there was still no way to know how many atoms a molecule contained. Some gaseous elements have diatomic molecules. Others have molecules with six or eight atoms. Without knowing the number of atoms in a molecule, Avogadro could not calculate atomic weights or formulas.

More important, however, Avogadro contradicted accepted theory without offering anything to take its place. He denied that atoms of an element repel each other. Instead, he asserted that atoms attract each other to form molecules. On the other hand, his molecules (for whose existence he had no proof) did not attract each other and perhaps even repelled each other. Why atoms should attract and molecules repel, he could not say. He had not even a suggestion of an explanation or a mechanism. It took more than a century, until 1919, for the first explanations to emerge, when Gilbert N. Lewis, Irving Langmuir, and Walter Kossel worked out the electron pair bond. No wonder that in 1811 Avogadro's contemporaries were skeptical.

The sticking point was the concept of attractions between atoms. Not only was there no evidence for interatomic attraction, but there was plenty of evidence for interatomic repulsion. The great Newton had used repulsion to explain gas pressures. Interatomic repulsion was also the theoretical basis of Dalton's Law of Partial Pressures, which had been experimentally verified. Moreover, Dalton's calculations of atomic weights and molecular formulas had been predicated on the idea that like atoms repelled each other. To complete the picture, there seemed to be an electrical explanation for interatomic repulsions. The early electrochemist Sir Humphry Davy and the analytical chemist Berzelius had developed a theory of chemical bonding independent of Dalton's theory: All atoms of a given element had the same electrical charge and should repel each other.

The Berzelius–Davy Bonding Theory

After the invention of the electrical storage battery in 1800, within ten years, Davy had isolated sodium, calcium, potassium, magnesium, and barium by electrolyzing molten salts. He observed that these metals, as well as hydrogen, appeared at the negative pole of the battery and decided that they must all be positively charged, because they are attracted by the negative electrode. In 1818, combining this idea with his own experimental work, Berzelius developed an electrochemical theory of bonding. All elements were supposed to be either positively or negatively charged. Positives repelled positives. Negatives repelled negatives. Only positive and negative elements, which attracted each other, could combine.

This first general theory of chemical bonding was accepted enthusiastically by chemists and physicists alike. It united chemistry and electricity in one generalization, and it made sense. It still makes sense for the broad class of materials that we call ionic compounds. Unfortunately, it does not apply to the much larger class of compounds that are covalent, including gases. (In ionic compounds, electrons are transferred from one atom to another. In covalent compounds, electrons are shared between atoms.)

To anyone who believed in the Berzelius–Davy idea of electrochemical bonding, molecules could not possibly be composed of several atoms of any one element. All hydrogen atoms were positive and therefore would certainly repel each other. Plainly, hydrogen atoms could not possibly form diatomic molecules. Nor could oxygen or chlorine form diatomic molecules. All oxygens and all chlorines were negative and would repel each other. Avogadro could not possibly be right.

Dumas's Molecular Weights

But, to some chemists, Avogadro's hypothesis seemed sufficiently useful for them to compromise and consider (correctly, as it turned out) that the Berzelius theory just might not hold for gases. In that case, they could apply the Avogadro hypothesis to gases. One of these compromisers was the brilliant young Jean-Baptiste André Dumas (1800–1884). In 1826, he developed the Dumas method for determining the densities of vapors (used in college chemistry courses as late as the 1960s). From his results he calculated the molecular weights of his vapors. Then, assuming that all gas molecules are diatomic, he calculated atomic and molecular weights. To his dismay, his atomic weight of phosphorus was twice the accepted value, that of sulfur was three times the accepted value, and that of mercury was half the accepted value. At that stage, he wished to remove the word *atom* from the chemist's lexicon because obviously it was being used for particles that contained several atoms (*8*).

Dumas nevertheless made possible the triumph of Avogadro. In 1833 he showed that in organic compounds the supposedly negative element, chlorine, could replace the supposedly positive element, hydrogen. Therefore, hydrogen and chlorine could not have opposite charges. In fact, they probably were not electrically charged at all. This was the beginning of the end for the Berzelius–Davy electrochemical theory of bonding, and by 1842 it was dead. By then, however, there was so much confusion that some chemists began to wonder if atoms really existed. There were almost as many different sets of atomic weights as there were chemists, and for every organic compound, there were dozens of different formulas, based upon the different atomic and molecular weights.

However, from about 1837 to 1863, Laurent, Gerhardt, Williamson, Wurtz, Couper, Kekulé, and others worked out many correct formulas. Furthermore, the Caloric theory of heat, Dalton's original justification for the concept of interatomic repulsion, was in decay and by 1857 had been replaced by the Kinetic Molecular theory. Consequently, there was no longer any argument against the existence of polyatomic molecules of elements.

At last, in 1860, an international conference of chemists was convened to discuss the problem of atomic weights. After three days of argument, the conference participants got nowhere, and finally everyone went home. On the last day, however, an Italian chemist named Pavesi distributed copies of a paper based on a short set of notes that Stanislao Cannizzaro (1826–1910) had been distributing to his students since 1858. In this pamphlet, *Sunto di un Corso di Filosofia Chimica* ("Epitome of a Course in Chemical Philosophy"), Cannizzaro explained and justified Avogadro's hypothesis, and showed how correct atomic weights and formulas could be obtained by easy calculations. His method was based on the simple idea that when equal volumes of different gaseous compounds are analyzed for a given element, the calculated weights of that element are found to be multiples of a common factor. The largest common factor is the atomic weight. Analysis of equal volumes of gases for all other elements results in a table of accurate relative atomic weights.

Many, but by no means all, of the chemists at the meeting read his work and were convinced. To quote Lothar Meyer, "The scales seemed to fall from my eyes." The Cannizzaro method rapidly became the accepted way of determining atomic weights and obtaining correct chemical formulas. Still there were hold-outs, and as late as 1870, many chemists, especially in conservative England, used HO as the formula for water.

References

1. Thackray, A. *John Dalton, Critical Assessments of His Life and Science*; Harvard University Press: Cambridge, MA, 1972; p 11 et seq.
2. Partington, J. R. *History of Chemistry;* Macmillan: London, New York, 1962; Vol. III, pp 772–773.
3. Nash, L. K. *Isis* **1956,** *47,* 1001–1016.
4. Partington, J. R., op. cit, p 784, principle ix.

5. Partington, J. R., op. cit., pp 783–786.
6. *Mem. Soc. Arcueil* **1809,** 207–234 and *Nuov. Bull. Soc. Philomath.* **1808,** *i*; 298, cited in Partington, J. R., op. cit., Vol. IV, p 79.
7. Partington, J. R., op. cit., Vol. IV, p 82.
8. Rocke, A. *Isis* **1978,** *69,* 595–600.

XIV

Atomic Arrangements

Organic Chemistry

THE EARLY NINETEENTH CENTURY CHEMISTS did not simply fill in the broad outlines of Lavoisier's new chemistry. There were no outlines. There was agreement on a useful nomenclature, on the basic concepts of elements and chemical reactions, and on the utility of formulas. But there was no program, no grand design. There were only individual chemists working on interesting problems, trying out ideas, running reactions that might or might not work. They made a large number of observations, some of which were right and some wrong. Somehow, the contradictions cancelled each other out, and a coherent theory remained. In tracing the course of events, we find there were fruitful errors, correct predictions based on reasons that seemed right at the time but that turned out to be wrong, and, in hindsight, misinterpretations.

At the start of the century, chemists generally agreed that the properties of substances depend on the composition of the compound atoms, or molecules, as they are called today, but no one knew anything yet about molecules. No one even knew how to go about investigating them. In France in 1784 the Abbé René J. Haüy proposed that the structure of crystals depended entirely on their constituent molecules. He maintained that it was the function of the crystallographer to determine from the structure of the crystal "the precise form of its constituent molecules, their respective arrangements," although oddly enough he remained skeptical about the physical existence of those constituent molecules (*1*). In spite of Haüy's vision, crystallography furnished very little information about molecular structure before the work of Louis Pasteur. Most developments took place in organic chemistry, the chemistry of covalent molecules.

It took half a century of confusion, bitterness, and even tragedy to prove that each molecule consists of atoms arranged in a definite pattern, to show that the properties of each substance depend not only on the number and kind of atoms present but also on the way atoms are arranged within the molecule, and then to work out some of these arrangements. Only when some structures were known was it possible to correlate molecular structure, at least empirically, with physical and chemical behavior of substances. Then, at last, chemists could begin to synthesize molecules to serve definite purposes.

The Beginnings

In 1800 organic chemistry was in a relatively primitive state. No one could say how organic chemicals differed from inorganic chemicals. Some believed that organic chemicals needed a "vital force" and could not be made in the laboratory. The comparatively few organic compounds known were natural products, such as waxes, fats, oils, acetone, sugars, oxalic acid, urea, and alcohol. The obvious first step was to purify and analyze the available organic compounds to see what they had in common. Then, perhaps, some generalizations might be made.

The first systematic studies of organic composition were those reported by Lavoisier in 1786 (*2*). On the basis of his own studies and those of Claude Berthollet (*3*), he defined organic chemicals as combinations of oxygen with radicals that contained carbon and hydrogen. If the organic compound was of animal origin, it also contained nitrogen and phosphorus. He used the combustion method of analysis, which, with modifications, was the standard for determining the quantitative composition of organic substances and was in use for over a century and a half. He weighed out samples of his substances, burned them in an atmosphere of oxygen, and trapped and weighed the carbon dioxide evolved. From the weight of carbon dioxide, he could calculate the percentage of carbon in the original material. Unfortunately, not only did he have no way to calculate the percentages of any other element, but even his analysis for carbon was not really accurate because sometimes the sample did not burn completely.

The first improvements came from Gay-Lussac and Louis Thenard (1777–1857) in 1810. They oxidized with solid potassium chlorate instead of oxygen and also analyzed for elements other than carbon. Berzelius made further refinements starting in 1814. He improved both precision and accuracy, and analyzed hundreds of the new compounds that were just beginning to appear.

Berzelius not only made quantitative analyses of substances, but also tried to calculate their equivalent weights and molecular formulas. In working out these molecular formulas, he compromised and did not apply his own theory of electrochemical bonding to gaseous elements (*see* Chapter XIII). Instead he assumed that, for elementary gases such as hydrogen, oxygen, and nitrogen, the ratio of the numbers of atoms was the same as the ratio of the volumes of the gases. (By implication, therefore, Avogadro was correct about elementary gases.) Berzelius was able to work out many correct formulas, but he made errors that had far-reaching consequences for the next thirty or more years.

In dealing with compounds that were not gases, Berzelius fell back on Dalton's rule of greatest simplicity. He assumed that, unless there was strong evidence to the contrary, atoms and radicals combined with each other on a one-to-one basis. This led to trouble. With divalent metals such as calcium and magnesium, his assumption worked out well, but for monovalent and trivalent atoms, his results were very wrong. For example, his formula for silver oxide was AgO instead of Ag_2O. The result was that he got twice the correct value for the atomic weight of silver. And when he used this value in calculating molecular weights of silver salts of organic acids, again he got doubled weights and therefore doubled formulas.

In spite of the confusion, knowledge accumulated. Known compounds were being analyzed, new substances and reactions discovered. Chemists were groping toward some sort of understanding of organic chemistry. Still there were obstacles. Lavoisier had defined organic compounds as radicals combined with oxygen, but he did not, could not, specify what a radical was. For a while, chemists thought that waxes and oils did not contain oxygen and that therefore they might be Lavoisier's radicals. Michel Chevreul (1786–1889), however, soon showed that waxes and oils did contain oxygen and so were not radicals, at least by Lavoisier's definition.

Chevreul did extensive work on oils, fats, and soaps between 1810 and 1823. He used inert solvents to extract chemicals, separated fatty acids from each other by recrystallization, and used melting points as criteria of purity. These techniques rapidly became standard. His results showed that oils and fats were combinations of organic acids with glycerol, and that all soaps were salts of organic acids. This was probably the first time that the chemical composition and behavior of entire classes of organic compounds had been worked out, and his work became the prototype for many other studies. However, although Chevreul clarified the composition of waxes and oils, the nature of radicals still remained up in the air.

Another difficulty in understanding organic chemistry was the vital-force concept. If organic chemicals could not be made in the laboratory and needed a vital force, they might not obey the same laws as inorganic chemicals (*4*). In that case, what laws did they obey?

Between 1811 and 1817, Berzelius, who formulated the first system of organic chemistry, analyzed enough organic compounds to obtain a significant number of experimental weight–weight relationships. On the basis of these, he decided that, although organic chemicals could not be synthesized, they did obey the accepted laws of chemical combination even though they often consisted of large numbers of atoms, and their formulas were complicated. He then applied his Dualistic Electrochemical theory to the internal binding of organic compounds (*see* Chapter XIII). Combining that theory with Lavoisier's ideas, he proposed what was later called the Older Radical theory. The name is confusing because there was no New Radical theory. He postulated that organic substances are oxides of compound radicals that usually consist of carbon and hydrogen and sometimes nitrogen, and are held together by electrostatic forces. Moreover, these radicals were stable groupings that could appear also in the free state, like stable chemical compounds. This valuable set of ideas was a working hypothesis that inspired further studies and led to significant advances.

Meanwhile, some very significant information was coming to light. In 1811 Gay-Lussac and Thenard analyzed samples of purified sugar, gum, starch, resin, wax, olive oil, and five organic acids. They found that in sugars, starches, woods, and gums, hydrogen and oxygen were present in the same proportions as in water. Apparently all these materials were hydrates of carbon, or "carbohydrates." In 1814 Gay-Lussac noted that both acetic acid and cellulose had the same composition: three grams of water to two grams of carbon (5). These two very different chemicals not only contained the same elements, but contained them in the same proportion. Obviously, as Gay-Lussac clearly stated in this important paper, the properties of organic molecules are determined not only by the number and kind of atoms present, but also by the arrangement of atoms (5). There had to be some kind of internal structure. Then, in 1815, Gay-Lussac apparently confirmed the Radical theory with the isolation and study of the gas cyanogen. He found that cyanogen underwent a whole series of reactions with organic compounds, in each of which the reaction product contained a CN group. For example, cyanogen and potassium hydroxide reacted to produce KCN. He therefore reasoned that cyanogen was the stable radical, CN, and that it maintained its identity during reactions. (Actually, the cyanogen molecule is N–C–C–N, CN being only the empirical formula.)

Meanwhile, more discoveries were being made. In 1816 Gay-Lussac found that there were two varieties of stannic acid with the same analytical composition but with different properties. At first this was regarded only as a puzzling aberration. Then, in 1823, Justus von Liebig (1803–1873) published the analysis of silver fulminate, and in 1824 Friedrich Wöhler (1800–1882) published that of silver cyanate. These two completely different compounds had the identical composition: the same percentages of silver, carbon, nitrogen, and oxygen. Gay-Lussac at once interpreted this as further evidence that atoms combine with each other in several different arrangements.

Not everyone accepted his interpretation: Berzelius, for example. After Liebig and Wöhler had rechecked their results to make sure they were accurate, Berzelius in 1827 explained away the difference in properties between silver cyanate and silver fulminate by proposing that it was the result of different vital forces. But in 1828 Wöhler dealt a blow to the vital-force concept. He heated the inorganic compound ammonium cyanate and got the organic substance, urea. Wöhler had actually prepared an organic compound in the laboratory without involving any vital forces (or convincing many vitalists, for that matter[1]). Moreover, ammonium cyanate, NH_4CNO, and urea, NH_2CONH_2, had identical percentages of carbon, hydrogen, nitrogen, and oxygen. It was another case of different compounds having the same

[1]Even after Wöhler's synthesis of urea, it took years to prove conclusively that vital forces were not present in organic chemicals. Adherents maintained that the ammonium cyanate could have contained the vital forces. Finally, both Kolbe, in 1845, and Berthelot synthesized organic compounds directly from the elements.

composition. Soon the phenomenon was found to be widespread, and in 1831 Berzelius coined the term *isomerism* for it.

By now chemists generally accepted the fact that organic radicals did have some sort of structure, but exactly what was meant by structure? No one was really sure. To most chemists, atoms somehow held together in a vague clump. A few had more exact ideas. In 1823 Chevreul wrote, "Species of compound bodies are identical when the nature, proportion, and arrangement of the elements are the same." In 1844 Eilhard Mitscherlich (1794–1863), in reporting that sodium ammonium tartrate was optically active and sodium ammonium paratartrate was not, wrote of the two compounds that "the nature, number, arrangement, and distances of the atoms" were completely identical (*6*).

Confusion over Atomic Weights and Formulas

In any case, whatever the term *structure* meant, to get a better understanding of radicals it was necessary to know their formulas. But how could chemists investigate the formula of a radical if they could not isolate and analyze it? Moreover, even if it could somehow be chemically analyzed, there were major problems in determining formulas.

To begin with, to calculate formulas from analytical data, it was necessary to know accurate atomic and molecular weights. But in the 1820s and 1830s there was almost total confusion over atomic and molecular weights and volumes. Different workers used different values of atomic weights, and therefore calculated different formulas from the same analytical data. Both Liebig and Dumas used 6 as the atomic weight of carbon, and their formulas showed twice as many carbon atoms as of those who used 12. Others used 8 for oxygen, instead of 16, so their formulas showed twice as many oxygen atoms.

A second source of difficulty was disagreement over the molecular volumes of gases. Berzelius calculated the molecular weights of gases the way we do. He assumed that the volume of a molecular weight of vapor was the same as that of two grams of hydrogen, that is, that hydrogen was H_2. However, Liebig and, for a time, Gerhardt based molecular weights on the volume of four grams of hydrogen, the so-called four-volume formulas. Liebig's gaseous molecular weights were, therefore, twice those of Berzelius, and, because he also used 6 as the atomic weight of carbon, they also showed four times as many carbon atoms. Dumas did the same, reporting methane as C_4H_8 instead of our CH_4. Mitscherlich, on the other hand, used one-volume formulas, and so reported benzene as C_3H_3 instead of our C_6H_6. Laurent reported naphthalene as $C_{40}H_{16}$ instead of $C_{10}H_8$.

Berzelius also assumed that gases such as hydrogen chloride and oxygen were evolved as double atoms, for example, $(HCl)_2$ not HCl. That meant any organic molecule that gave off hydrogen chloride in the course of a reaction had to contain at least two chlorine atoms. In effect, Berzelius doubled the molecular weight and molecular formulas of organic chlorides. Other organic molecules that gave off gases had their molecular weights and formulas doubled, but molecules that did not give off gases did not have their weights doubled. Some did; some didn't. Chemistry was in almost total confusion.

Another set of problems arose from Lavoisier's original concept of an acid. Oxygen was the acidifying principle central to his theories. Muriatic acid, our HCl, was a problem to his followers. It contained no oxygen. As early as 1813 Davy had noted that acids contain hydrogen and suggested that hydrogen, not oxygen, was the essential ingredient in acids. Except in England, however, chemists preferred Lavoisier's definition. They explained away the awkward hydrogen found in the acid by calling it part of the "water of composition." Whenever an acid was found to contain hydrogen, the elements of water of composition were subtracted from the formula. For example, to eliminate the hydrogen from H_3PO_4, H_2O would be subtracted, leaving HPO_3. But this still had one hydrogen and so, to get rid of that, its formula was doubled to $H_2P_2O_6$ and H_2O subtracted again, leaving P_2O_5 as the accepted formula for phosphoric acid. This is not phosphoric acid but phosphoric acid anhydride, the oxide that combines with water to form phosphoric acid.

When Berzelius began analyzing organic acids, he, too, eliminated hydrogen, so his formulas for acids were also those of anhydrides. For example, he assigned acetic acid the formula $C_4H_6O_3$ (instead of $C_2H_4O_2$). Ethanol was known to form acetic acid on oxidation, but Berzelius correctly assigned it the formula C_2H_6O, and so the structural relationship between ethanol and acetic acid vanished. Liebig, however, had decided that ethanol was $C_4H_{12}O_2$ in order to represent it as a hydrate of ether. In 1834 Berzelius came around to this formula, because ether is prepared by dehydrating ethanol.

The Etherin Theory

In the midst of all this confusion, Dumas made a start on a theory of molecular structure on the basis of work originally done by

Gay-Lussac in 1814–1815. Gay-Lussac had observed, without understanding why, that the density of alcohol vapor was equal to the sum of the density of water vapor plus that of ethylene vapor. Additional work showed that ethyl chloride had a density equal to the sum of the densities of ethylene and hydrogen chloride. As we now know, the densities of vapors and gases are proportional to their molecular weights, which means that the molecular weight of ethyl chloride is the sum of the molecular weights of ethylene and HCl. Assuming this without being able to prove it, in 1828 Dumas and Polydore Boullay (1806–1835) proposed the Etherin theory, as Berzelius named it in 1832. (In effect, they were accepting Avogadro's hypothesis.) According to Dumas and Boullay, ethylene was a radical, called etherin. Ethyl alcohol, ethyl ether, ethyl acetate, ethyl chloride, and ethyl sulfate were all products of the addition of one or more molecules to ethylene. For example, using our formulas:

$$\underset{\text{(ethylene)}}{C_2H_4} + \underset{\text{(water)}}{H_2O} = \underset{\text{(ethyl alcohol)}}{C_2H_6O}$$

The Etherin theory, although inadequate, foreshadowed the future. It was the first theory that showed structural relationships between different compounds, which suggested that some organic compounds were derived from others.

In that same year, 1828, Liebig and Wöhler proposed a different radical, the benzoyl radical. From oil of almonds they had produced benzaldehyde, and starting with benzaldehyde they were able to prepare benzoic acid, benzoyl chloride, benzoyl bromide, benzoyl iodide, benzoyl cyanide, benzamide, and ethyl benzoate. In all these compounds, they found analytical evidence for a characteristic group to which they assigned the composition $C_{14}H_{10}O_2$. They decided that this group was a radical, gave it the name benzoyl, and concluded that all the compounds they had formed were addition products of the benzoyl radical.

At first, Berzelius could not have been more pleased. Benzoyl contained oxygen. It therefore proved that Lavoisier was right. By 1833, however, he had rethought his position. Radicals should be groups that combined with oxygen, not groups that contained oxygen. He now suggested that the benzoyl group was not really a radical, but merely an oxide of $C_{14}H_{10}$, the true radical.

Meanwhile Robert Kane (1809–1890) proposed still another radical, the ethyl radical, C_4H_5 (our C_2H_5). To Kane, ethyl, not ethylene, was the radical in the etherin series of Dumas and Boullay. At first Liebig accepted this ethyl radical, but in 1839 he proposed what he considered to be a better unifying idea, the acetyl radical, C_4H_6 (our C_2H_3). If hydrogen were added to acetyl, one would get ethylene. If more hydrogen were added, one would get ethyl, and if oxygen and hydrogen were both added, one would get acetic acid.

Little by little, between the time that cyanogen and ethylene were first thought to be radicals and the time that Liebig suggested the acetyl group as a radical, the meaning of the term had changed. Originally, a radical was supposed to be a real compound, capable of independent existence. Ethylene was a real substance, and so was cyanogen. However, by 1839, to Liebig the term *radical* obviously meant an arrangement of atoms that might be completely hypothetical, such as acetyl, an arrangement that was common to a series of compounds but that had no independent existence.

By now there were serious problems with Berzelius's theory that all bonds between atoms were the result of electrical attractions, with every pair of bonded atoms having opposite charges. Chlorine, bromine, and oxygen were supposed to be negative, but hydrogen and the metals were positive. In 1823 Gay-Lussac had called attention to the fact that in cyanogen chloride, chlorine replaced hydrogen, but chemists had not yet realized the significance of this.

Then in the early 1830s, Dumas and Auguste Laurent (1808–1853) pulled the props from under the entire concept of electrostatic bonding in organic molecules. First, in 1833 Dumas reported that in some hydrocarbons the negative chlorine could replace the positive hydrogen. In 1834 he demonstrated that both the positive hydrogen and the negative oxygen could be replaced by the same atom, chlorine. He prepared a whole series of examples, in many of which substituting chlorine for hydrogen had not significantly changed the chemical properties. Then Laurent produced further evidence against the electrochemical theory by reporting that he had nitrated and dinitrated naphthalene, with the negative nitro group replacing the positive hydrogen. Initially, there was considerable skepticism about Laurent's results, especially on the part of Liebig and Berzelius (Berzelius was, after all, one of the authors of the electrochemical theory of bonding). They refused to believe Laurent and made violent personal attacks on him. Liebig was particularly abusive, calling Dumas a "highway robber"[2]. Nevertheless, between 1834

[2]In the nineteenth century, chemists editing journals often made sarcastic attacks on the work and character of their opponents, even inserting derogatory comments into the author's article. Liebig wrote that Dumas stole other scientists' research. He also declared Laurent ignorant "of the principles of scientific research" (7).

and 1838 so many substances were substituted that the evidence became overwhelming. Then in 1839 Dumas produced trichloroacetic acid and showed that its properties were similar to those of acetic acid. Finally, in 1842, Louis Melsens (1814–1886) converted trichloracetic acid back into acetic acid by replacing all the chlorine with hydrogen. That was pretty much the end of Berzelius's electrochemical theory of bonding, at least for organic molecules.

It is not hard to understand why Berzelius and Liebig clung to the electrochemical theory. If molecules are not held together by opposite electrical charges, what holds them together? In 1840 there simply was no answer. Except for the attraction of opposite charges, there was no other theory of bonding at all. Some chemists continued to believe in molecular structures whose bonding might someday be explained. Others, along with physicists, began to doubt the physical existence of atoms and molecules. Charles Gerhardt (1816–1856), for example, explicitly denied that the formulas he laboriously constructed represented the actual composition of a compound. To him, they were simply mental constructs that might be useful in pointing up the way molecules reacted. In fact, to Gerhardt, in different reactions the same compound might be represented by several different formulas. In discussing cases such as acetic acid and trichloroacetic acid, in which the two different molecules have similar properties but are composed of different combinations of atoms, he reasoned that the properties of the molecule could not depend on the structure because the structure should change as the atoms were changed.

Ironically, the two men who never lost their belief in the reality of atoms and molecules and of molecular structure were Dumas and Laurent, the two who had done the most to destroy Berzelius's electrochemical theory.

Auguste Laurent

Laurent, who had been Dumas's assistant and co-worker, was a brilliant scientist who did some remarkable work in spite of overwhelming hardships. He was a fine experimentalist who discovered, among other things, hydrobenzamide, phthalic anhydride, dichlorophenol, picric acid, and stilbene. He was also an outstanding theorist who outstripped his slower contemporaries, to his own detriment. They simply could not keep up with him and so took a long time to accept his ideas. He brought one problem on himself, that of nomenclature. He tried to revise the chemical nomenclature of his times on logical grounds by inventing new, and often confusing, names for familiar compounds. These were ignored. His personality also stood in his way. Although amiable and pleasant, he was jealous of his claims to priority, and he often got involved in needless controversies, antagonizing those who might otherwise have helped him. As an avowed republican in the France of Louis-Phillipe's monarchy and the empire of Louis-Napoleon, he needed all the help he could get. He was never able to obtain an appropriate laboratory position, and never able to earn a decent living for himself and his family.

For long periods Laurent eked out a bare livelihood, teaching private pupils, living a life of Mozartean poverty. There is a pathetic letter extant asking his friend Gerhardt for twenty francs (then about four dollars) to buy chemicals because he had no more alcohol or ether. Twice he applied for professorial positions, and twice they were given to a relative nonentity who already had two laboratories and was doing little with them. He spent his last years working in a cellar that he had converted into a laboratory. The intolerable conditions finally broke his health. He died of tuberculosis at the age of forty-five, leaving his wife and two children, one an infant, completely destitute. Friends in England and America helped the widow and, now that he was safely dead, his colleagues arranged for the government to give his family a small pension and to educate his children. Incidentally, this was a good investment, for his son became a famous mathematician.

Laurent's Nucleus Theory

In 1837 Laurent presented as his doctoral thesis a theory of organic substitution. It not only proposed that the properties of a compound depend on the arrangement of atoms, but actually set forth the idea of three-dimensional structure, the beginnings of modern stereochemistry. The radicals, or "nuclei," as he called them, had the shape of cubes and other three-dimensional figures. For example, the C_8H_{12} radical had eight corners, each occupied by a carbon atom, and twelve edges, the center of each edge containing a hydrogen atom. If atoms were *added* to or *removed* from the faces or edges, the geometry of the substance would be changed and so would its fundamental nature, which depended on the geometry. On the other hand, if a hydrogen atom were *replaced* by another atom, such as chlorine, the geometry and the properties would not change much. There is a great deal wrong with this theory, but it contains startling insights. Laurent distinguished between addition and substitution on the basis of stereochemistry; he recognized that the contribution of an atom to molecular properties depends on its position; and he implied that the basic properties of a chemical depend on its shape of the molecule.

Berzelius and Liebig immediately attacked Laurent's Nucleus theory, and brought Dumas into the controversy, because Laurent was his student and graduate assistant. Dumas uncharacteristically backed away from the fight, and protested that he had nothing to do with the Nucleus theory, claiming that Laurent had misinterpreted some of Dumas's experimental data. This mollified Berzelius and Liebig, but antagonized Laurent, who wrote to Dumas sarcastically thanking him for admitting that the credit for the Nucleus theory belonged to Laurent. From then on, the two were on bad terms, squabbling with each other, and always to Laurent's disadvantage, because Dumas was the most powerful man in the world of French chemistry and could have and should have helped him. Curiously, Dumas shortly thereafter brought out his own theory, the First Type theory, which differed only slightly from Laurent's.

Gerhardt's Collaboration

From about 1840 on, Laurent was supported professionally and aided financially by the young Charles Gerhardt, his collaborator and staunch friend. Gerhardt, too, was a republican who suffered political reprisals for his part in the Revolution of 1848, which overthrew the monarchy only to pave the way for Napoleon III. He, too, died young, just when his work had gained acceptance. Gerhardt was brash, rude, and tactless and needlessly antagonized even his own supporters. The combination of Laurent's revolutionary ideas and Gerhardt's tactlessness made the two men virtual outcasts. Not only did they challenge the viewpoints and vested interests of the very conservative French establishment, but they did so abrasively, forcing confrontations that they were bound to lose. Under the French system, the professor was an independent feudal magnate; the few professors controlled appointments, awards, and salaries.

In analyzing the collaboration of Gerhardt and Laurent, it is difficult to assess the separate contributions of each. Most of the ideas were probably Laurent's, and most of their development was that of Gerhardt, who wrote more clearly and whose articles were easier to understand. He was also an exceptional experimentalist who did important work on quinoline and tartaric acid and its derivatives, and discovered styrene, acetanilide, sulfanilic acid, and the monobasic anhydrides. An outstanding theoretician, he organized classification after classification until finally he succeeded in bringing some order into the confusion of organic chemistry.

In 1839 Gerhardt proposed a new theory of organic reactions called the Theory of Residues. Essentially, this amounted to the proposal that organic reactions proceed by double decomposition. At first glance the idea was attractive, but there are simply too many reactions that are not double decompositions, so the theory fell by the wayside. But Gerhardt persisted in trying to establish generalizations.

Starting in 1842, he published an important series of papers on molecular weight and structural classification, developing the principle of homology. According to this principle, in a homologous series the properties are similar and sometimes almost identical. For example, the acid consisting of a three-carbon chain ending in an acid group has, in general, the same chemical properties as the acid with a four-carbon chain and the one with a five-carbon chain. The difference in chain length produces some quantitative differences in properties, but the qualitative behavior depends primarily on the acid group at the end of the chain. Therefore, learning the properties of one acid gives insight into the properties of all acids. Carried away by enthusiasm, Gerhardt proclaimed, "It is only necessary to know the reactions of any one [in the series] in order to predict those of the others." This is, of course, quite an overstatement.

Next he turned his attention to a situation that outraged his logical mind. Clearly two different molecular weights should not be listed for the same substance. But for many vapors, the molecular weights used by inorganic chemists were half of those assigned by organic chemists. This was intolerable! The inconsistency had arisen because inorganic chemists assumed that water and other gases were evolved as single molecules. Organic chemists, on the other hand, used formulas based on the amount of material that would absorb or evolve double molecules.

In 1843 Gerhardt proposed his solution: In effect, Berzelius was wrong. Gases are evolved or absorbed as single molecules, and organic chemists should calculate molecular weights and formulas on that basis, bringing their results into line with those of inorganic chemists. He suggested that Berzelius's atomic weight of silver was wrong and proposed cutting it in half. This would cut molecular weights and formulas of organic acids in half, bringing them down to values that we now consider correct. Both suggestions would have resulted in better weights and formulas had they been accepted. But they were not. For one thing, Gerhardt still had not completely solved the problem.

Bringing organic formulas into line with inorganic formulas was not a cure-all, because many of the inorganic formulas and weights were themselves wrong. They were based on equivalent weights, but the equivalent weight of a substance is not necessarily its molecular weight.

Laurent, however, soon suggested a solution. In 1846 he published a very important article clearly explaining the difference between molecular weight and combining weight (*8*). The genesis of this work was a paper by Thomas Graham (1805–1869) on the phosphates of soda. Graham reported that there were several different phosphoric acids, the difference between them being the relative quantities of hydrogen, although acids were not supposed to contain hydrogen. Liebig picked up Graham's work and began studying molecules of organic acids, this time not disregarding the hydrogen present. Finally, he concluded that Davy had been right all along and that hydrogen, not oxygen, was responsible for the characteristic properties of acids. Now he recalculated the formulas and molecular weights of organic acids. In so doing, he discovered that some organic acids were polybasic: they could react with several molecules of base. This was the breakthrough that Laurent needed.

Laurent seized on the significance of Liebig's work and pushed it to its logical conclusion. If one molecule of a dibasic acid can react with two molecules of base, the equivalent weight of the acid is only half of the molecular weight. Equivalent weights and molecular weights are therefore different entities. Laurent went on to distinguish clearly between atomic, molecular, and equivalent weights. He defined the atom as the smallest unit of an element and the molecule as the smallest unit of a compound. Atomic weight is the smallest weight of an element present in a compound. (This idea led directly to Canizzarro's method of determining atomic weights.) The molecular weight is the total of all the atomic weights in the compound. The equivalent weight is not necessarily the molecular weight. It is just the weight of the compound that can take part in a given reaction, and it varies from reaction to reaction. Therefore, because the chemical equivalent weight of a substance is variable, molecular weights cannot be determined chemically. They must be found by physical magnitudes, such as vapor densities.

Applying these ideas, Laurent showed that elementary gases such as oxygen and nitrogen consist of diatomic molecules. He then published the following table, taken from his article in *Annales de Chimie et de Physique*, in 1846.

Each substance in the table is essentially a water molecule with either hydrogen or oxygen or both substituted by various

	ACIDE hydrique.	ACIDE sulfhydrique.	ACIDE sulfureux.	ACIDE sulfurique.	ACIDE carbonique.	ACIDE oxalique.	ACIDE camphorique.
Acide............	O HH	S HH	SO^3HH	SO^4HH	CO^3HH	C^2O^4HH	$C^{10}H^{14}O^4$,HH
Sel acide..........	O HK	S HK	SO^3HK	SO^4HK	CO^3HK	C^2O^4HK	$C^{10}H^{14}O^4$,HK
Sel neutre........	O KK	S KK	SO^3KK	SO^4KK	CO^3KK	C^2O^4KK	$C^{10}H^{14}O^4$,KK
Sel double........	O KM	S KM	SO^3KM	SO^4KM	CO^3KM	C^2O^4KM	$C^{10}H^{14}O^4$, KM
Hybrides..........	$O(K^{\frac{1}{3}}Al^{\frac{2}{3}}$	S (Cu, F, Zn)^2	"	SO^4 (Cu, N, F)^2	CO^3 (Cu, Mg, F)^2	$C^2O^4K^{\frac{1}{3}}H^{\frac{2}{3}}$	"
Acides viniques....	O EtH	S EtH	SO^3EtH (*)	SO^4EtH	CO^3EtH	C^2O^4EtH	$C^{10}H^{14}O^4$, EtH
Sels..............	O EtK	S EtK	SO^3EtK	SO^4EtK	CO^3FeK	C^2O^4EtK	$C^{10}H^{14}O^4$,EtK
Éthers............	O EtEt	S EtEt	SO^3EtEt	SO^4EtEt	CO^3EtEt	C^2O^4EtEt	$C^{10}H^{14}O^4$,EtEt
Anhydrides...... ..	"	"	SO^2	SO^3	CO^2	"	$C^{10}H^{14}O^3$
Acides amidés.....	"	"	SO^2H^2N	SO^3H^2N	CO^2H^2N	$C^2O^3H^2N$	$C^{10}H^{14}O^3A^2N$
Ammons...........	"	"	$SO^2H^4N^2$	$SO^2H^4N^2$	$CO^2H^4N^2$	$C^2O^2H^4N^2$	$C^{10}H^{14}O^2H^4N^2$
Anilons...........	"	"	"	SO^2An^2	CO^2An^2	$C^2O^2An^2$	"

(*) Les acides méthylhyposulfurique, chlorélaylhyposulfurique, chloroformylhyposulfurique, chlorocarbohyposulfurique de M. Kolbe (*métholates sulfurés,* GERH.), ne sont évidemment que les acides sulfométhyleux chlorés, bichlorés. De même, l'acide sulfosulféthylique est l'*acide sulfovineux,* et l'acide de M. Gerathewohl (par NHO^6 sur le mercaptan amylique) est l'*acide sulfoamyleux.*

Laurent's 1846 table showing formulas of various compounds of the water type. The first columns show formulas of water molecules with hydrogen being replaced by various atoms or groups. The other columns show the same substitution products but with the oxygen of the water molecules replaced by atoms or groups. Ethyl alcohol is under the heading of vinic acid. (Reproduced with permission from ref. 8).

atoms and groups (except, of course, for water itself). This logical grouping immediately demonstrated unsuspected or unnoticed relationships between substances, both organic and inorganic. In fact, the very first column shows clearly the relationship between alcohol and ether that Alexander Williamson, to his surprise, observed four years later (*see* next section).

Chemists rapidly became more comfortable with both the reality of structure and the idea of basic types of compounds, and support for Laurent and Gerhardt began to grow. At this stage the contributions of other young chemists became important.

Type Compounds

Charles A. Wurtz (1817–1884) was a classmate and a friend of Gerhardt who had the courage to stick by him and to teach his unpopular views. In 1849 Wurtz discovered ethylamine and methylamine and suggested immediately that these new compounds were molecules of ammonia with one hydrogen atom replaced by ethyl or methyl. Following up on Wurtz's work, August von Hofmann (1818–1892) isolated secondary and tertiary amines and proposed an ammonia type. He suggested that amines be considered derivatives of ammonia with one, two, or three hydrogens being replaced by carbon chains:

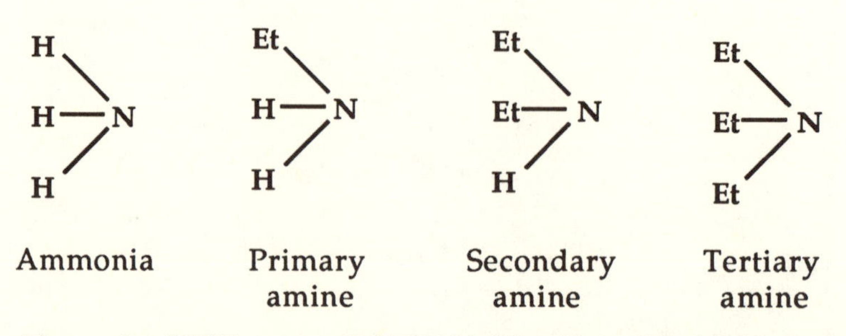

Alexander Williamson (1824–1904) had been Liebig's student and originally accepted his ideas. Although by 1850 he had changed his opinions and was now a firm supporter of Laurent and Gerhardt, he still retained Liebig's belief that alcohol was a hydrate of ether, containing both ether and water. The formula was $C_4H_{10}O \bullet H_2O$ with $C_4H_{10}O$ being the "copula," in which substitution should take place. (Berzelius in 1843 had reluctantly accepted the fact that substitution by atoms of different sign could take place. To save his Dualistic theory, he postulated that molecules were composed of two separate parts, but that substitution could occur in only one of them, the copula.) In 1850 Williamson set out to make higher alcohols by substituting radicals onto the copula (*9*). He started with potassium ethylate, which he wrote as $C_4H_{10}O \bullet KO$, and reacted it with ethyl iodide, C_2H_5I. He expected the reaction to be

$$C_4H_{10}O \bullet KO + C_2H_5I \rightarrow KI + C_6H_{15}O \bullet H_2O$$

(reaction not balanced here)

To his surprise Williamson obtained ordinary ether, $C_4H_{10}O$. The fact that his product was not an alcohol meant that substitution had not taken place in the copula, that the H_2O position did not remain intact. So the Copula theory was wrong. (There still remained a possible mechanism that might have saved the theory, but Williamson disposed of it by making ethyl methyl ether from potassium ethylate and methyl iodide.) Moreover, it was agreed that ether had the formula $C_4H_{10}O$, so ethanol and ethyl iodide had to have a total of four carbon atoms between them. Ethyl iodide was C_2H_5I, so ethanol had to be C_2H_6O.

Williamson generalized his results, declaring that the alcohol molecule was a water molecule with one hydrogen replaced by a hydrocarbon radical, and that ether was a water molecule with both hydrogens replaced by hydrocarbon radicals. (For some reason, Williamson did not mention that Laurent had already published this. It may be that he had not seen, or perhaps even forgotten, Laurent's paper.) In 1851 Williamson went further and included acetic acid as a water-type molecule.

In 1853 Gerhardt put all this new information together and proposed a comprehensive theory of type compounds, called the Second Type theory, Dumas's theory of 1839 being the First Type theory. This Second Type theory led through Kekulé's work to modern organic chemistry. Gerhardt suggested that there were four basic types of organic compounds: the water type, the ammonia type, the hydrogen type, and the hydrogen chloride type. In these compounds, if an organic group replaced a hydrogen atom, the products would be an alcohol or an amine or a hydrocarbon or a chloride, respectively. He then went through the list of almost all known organic compounds to show that each one could be fitted into one or more of these types.

The new theory was so comprehensive and so convincing that even Liebig lessened his opposition to Gerhardt's ideas, which quickly won acceptance. Tragically, in that same year, 1853, Laurent died at the moment of success. In 1856 Gerhardt too died suddenly at age forty. Had those two young men lived longer they would no doubt have made many more contributions.

Synthetic Organic Chemistry

The Second Type theory gave insight into relationships between classes of organic chemicals, but chemists were still ignorant of the actual structures. Ethanol might indeed be

$$\begin{array}{l} H \\ | \\ O\text{—}C_2H_5 \end{array}$$

but nobody could say what the C_2H_5 structure really was. However, chemists now had models of organic compounds that suggested both new reactions and new molecules to try to make. This period starts the great age of synthetic organic chemistry whose pioneers were Herman Kolbe[3] (1818–1884), Marcellin Berthelot (1827–1907), and Edward Frankland (1825–1899). Ironically, as new compounds were synthesized on the basis of the Type theories, the various relationships became clearer, but the

[3]Kolbe, paradoxically, was a disciple of Berzelius and devoted much of his efforts to proving the Copula theory correct, at least partly because, as a passionate German nationalist, he disliked Gerhardt and hated all things French. For example, he attacked both Berthelot and his work in scathing terms. He commented of one paper that what was new was not true and what was true was not new. He also stated that Berthelot knew less about chemistry than a schoolboy (7). Berthelot's work was at least the equal of Kolbe's.

types themselves immediately lost their importance. For example, methylamine might be considered to be either methylated ammonia or ammoniated methane. Ethane might be ethyl hydride, derived from hydrogen, or it might be methylated methane. There was still no clear idea as to the structure of the organic radicals or groups in the type compounds, but some of the answers were not long in coming, worked out by Couper and by Kekulé and based on the concept of valence, as proposed by Frankland.

Frankland and Valence

Edward Frankland, an Englishman, was a friend and co-worker of Kolbe and an admirer of Berzelius. In 1848 he set out to prove that organic compounds really consisted of a copula, in which substitution took place, and a group in which substitution could not take place. His approach was to produce and isolate hydrocarbon radicals to show that they could exist independently, and then to combine them to show that they could couple. He did succeed in isolating what he thought was zinc methyl and zinc ethyl (actually the dimethyl and diethyl) and, in fact, he was the great pioneer in the early development of organometallic chemistry, but he ended up disproving what he had set out to prove.

Frankland started with zinc dimethyl, which he considered to be $Zn{\bullet}C_2H_6$. According to Berzelius's theory, the zinc atom was the copula and therefore should be capable of being oxidized to ZnO without changing the rest of the compound. To his surprise, Frankland was completely unable to oxidize the zinc even though he could oxidize arsenic in the analogous arsenic compound. He finally concluded that zinc dimethyl could not pick up an oxygen and that there really was no copula to which oxygen could add without disturbing the organic radical. In fact, he allowed that the zinc and the other organometallic compounds he had been studying were of the oxide type with organic radicals substituting for oxygen. To his probable initial discomfort, Frankland had proved Berzelius wrong and Gerhardt right.

Furthermore, in the course of his work he had been struck by the fact that there seemed to be an upper limit to the number of groups that each central atom could bind. He reported his findings as follows:

> When the formulae of inorganic chemical compounds are considered, even a superficial observer is struck with the general symmetry of their constitution; the compounds of nitrogen, phosphorus, antimony, and arsenic especially exhibit the ten-

> dency to form compounds containing three or five equivalents of other elements and it is in these proportions that their affinities are best satisfied . . . Without offering any hypothesis regarding the cause of this symmetrical grouping of atoms, it is sufficiently evident from the examples just given that such a tendency or law prevails and that, no matter what the character of the uniting atoms may be, the combining power of the attracting element, if I may be allowed the term, is always satisfied by the same number of these atoms. [*10*]

This is probably the first statement of the idea of valence, of a specific combining power of an atom. Frankland, however, went no further than making the observation that what we now call valence seemed to be a property of many elements. He made no attempt to use the concept to develop formulas. These were worked out, independently, at about the same time by Couper and by Kekulé.

Couper and Kekulé

The brilliant Archibald Scott Couper (1831–1892) was a tragic figure. He was an unstable, unfulfilled genius who by age twenty-seven had produced epoch-making but unrecognized discoveries and then collapsed into oblivion. He started out as a student in philosophy at the University of Edinburgh and only after obtaining his degree did he develop an interest in chemistry. In 1856 he went to Paris to work in Wurtz's laboratory, where almost immediately he synthesized two new compounds, bromobenzene and *para*-dibromobenzene. In 1858 he published an important paper on salicylic acid that contained the first modern structural formulas. In that same year, he wrote a paper titled "A New Chemical Theory" and asked Wurtz to present it to the French Academy (*11*). Applying Frankland's concept of valence, this paper proposed that carbon atoms could combine with other carbon atoms to form chains and that, when carbon did combine, it did so either with four equivalents of hydrogen, as in derivatives of carbon dioxide, or with two equivalents, as in derivatives of carbon monoxide. Couper considered that the formation of these carbon chains was the key to the structure of organic compounds. He stated

> I propose to consider the single element carbon. This body is found to have two highly distinguished characteristics: 1. It combines with equal numbers of equivalents of hydrogen, chlorine, oxygen, sulfur, etc. 2. It enters into chemical union with itself. These two properties in my opinion explain all that is characteristic of organic chemistry [*11*].

Wurtz was not at that time a member of the French Academy and delayed some months before giving the paper to someone who was a member. In the interim, Kekulé's paper appeared, also proposing that carbon combined with carbon and was tetravalent. As a result, Friedrich August Kekulé von Stradonitz (1829–1896) received the credit and Couper, rejected, disappointed, and dejected, returned to England, where shortly afterward he had a nervous breakdown. He never completely recovered, spending the rest of his life as a semi-invalid. He was completely forgotten until around 1900, when Richard Anschutz discovered his work and brought it to the attention of the scientific world. It was poetic justice. Anschutz had succeeded Kekulé as professor of chemistry at the University of Bonn, and ran into the first mention of Couper in the course of writing a biography of Kekulé.

Couper's papers on structural organic chemistry differed from Kekulé's 1858 paper in several significant details. His graphic formulas were better and clearer. He proposed a ring structure for cyanuric acid in 1858, whereas Kekulé did not arrive at his ring structure until 1865. Couper also recognized that carbon had two valences, while Kekulé believed carbon was always tetravalent. Finally, Couper advanced the revolutionary idea that the atoms themselves might be composite bodies forty years before the work of J. J. Thomson, while Kekulé doubted whether atoms exist. Couper insisted on the reality of molecules and their structure and was perhaps the first to define structure as the connection between atoms to form molecules. This contrasted with the attitude of Gerhardt, who regarded formulas as "a convenient means of representing the reactions in which a substance might take part" and that of Kekulé, who for many years considered that the concept of atoms was a "convenient fiction."

Here are some of Couper's formulas, using 8 as the atomic weight of oxygen, which required him to use two oxygen atoms where we draw only one:

. O• • • •OH
.
.
C• • • •H_3

Methanol

. O• • • •OH
.
.
C• • • •H_2
.
.
.
C• • • •H_3

Ethanol

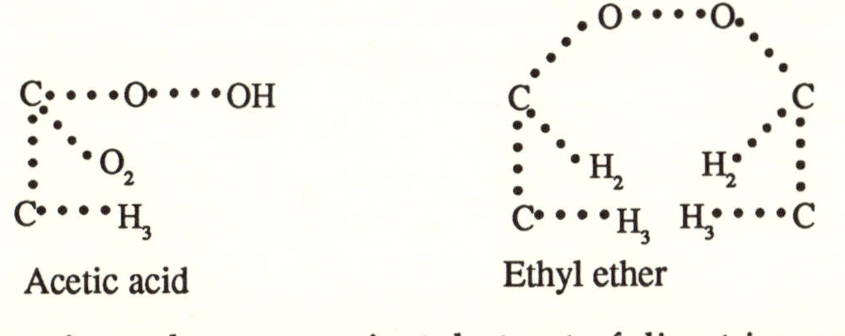

Acetic acid

Ethyl ether

Couper's work was prescient, but not of direct importance. Because of Kekulé's claim to priority and Couper's disappearance, organic chemists studied Kekulé's work. He must be considered the most important figure in establishing structural organic chemistry.

Kekulé had studied architecture, which in his work on molecular structure came in handy. He was a friend and follower of Gerhardt, but in his work on the structure of the hydrocarbon skeleton he left Gerhardt far behind. In 1858 he proposed that carbon was tetravalent, what we now call covalent, and in 1865 he proposed that benzene had a cyclic structure, with the ends of the chain bending around to form a hexagon[4]. He proposed that disubstituted derivatives of benzene should differ from each other if the substituting groups were in different positions. (This is the first mention of positional isomerism.) He was also a prime mover in convening the first international conference on chemistry, in Karlsruhe in 1860, at which Cannizzaro presented his method of calculating atomic and molecular weights. Kekulé's work met with immediate acceptance and the main lines of structural covalent chemistry were rapidly worked out, except for uncertainties about the nature of bonding. Kekulé and his followers found out *what* without the faintest idea of *why*.

Optical Activity and Pasteur

The one great development in structural chemistry that had not yet been considered was stereochemistry, the shape of the molecules. The start of stereochemistry was Laurent's Nucleus theory. It had been largely ignored, but in 1848 that first-rate young chemist Louis Pasteur (1822–1895) revived the idea of spatial

[4]Laurent had already shown benzene as a hexagon in his *Methode de Chemie*, published posthumously in 1854. Kekulé knew of this and even referred to the page in 1858 (*12*). However, even if it did influence his thinking, it made no great impression on him since he did not publish his benzene formula for another seven years. By then, he had probably forgotten Laurent's drawing.

chemistry. Pasteur was a friend of Laurent and for his doctorate worked on a problem suggested by Laurent, who was deeply interested in crystallography.

Laurent believed that similar molecules formed similar crystals. Sodium tartrate and potassium tartrate should have the same crystal form. Moreover, addition of water molecules should change crystal shape and the change would depend on the number of water molecules added. For example, sodium tartrate monohydrate and potassium tartrate monohydrate should have the same crystal structure and be isomorphic; that is, crystals of either could serve as nuclei for crystallization of the other, each new crystal containing both compounds. Furthermore, substances with the same crystal form should have the same optical activity. (In 1812 Jean B. Biot [1774–1862] had discovered that certain liquids, solutions, and crystalline solids were optically active; that is, when polarized light was passed through them, the plane of polarization was rotated.)

Laurent suggested to Pasteur that for his chemistry thesis he study the salts of tartaric acid to correlate their formulas with crystal structure and properties, especially optical activity. Pasteur took on the problem, expecting to substantiate Laurent's ideas. As a start, he searched the literature to see what was known about tartrates and soon ran into the peculiar case of sodium ammonium tartrate and sodium ammonium racemate, or paratartrate as it was also called. He found that Eilhard Mitscherlich had reported in 1844 that these two salts had the same chemical composition, the same crystal form, and identical chemical properties, but although the tartrate was optically active, the racemate was inactive. Optical activity should depend on crystal form; therefore, if Mitscherlich was right, Laurent was wrong. Pasteur had to check Mitscherlich's work.

He began by examining the crystals under the microscope, as others had done. His careful scrutiny showed him what they had missed, namely, that the crystals differed slightly. Both salts were hemihedral, but the tartrate crystals were all identical, while the racemate was a mixture of two kinds of crystals. The tartrate crystals were all at the same *angle* and direction, but half of the racemate crystals were oriented at an angle to the right, and the other half at the same angle to the left. Furthermore, those racemate crystals oriented to the right were crystallographically identical with the tartrate crystals.

Pasteur immediately surmised that the racemate was inactive only because the two sets of racemate crystals had opposite optical rotations that canceled each other out. To prove it, he separated one racemate set from the other (by picking the crystals up, one by one, with a pair of tweezers), and determined their optical rotations. Not only were they indeed equal and opposite, but the right-handed set had the same rotation as did the tartrate crystals. The racemate was therefore not a pure compound, but just a mixture of tartrate crystals with their mirror images. The problem of the racemates was solved. But in solving it, Pasteur had run into new problems that ultimately opened up a whole new area of research, that of stereochemistry.

Pasteur had been lucky because the starting racemic mixture of sodium ammonium tartrate crystallizes into two different forms only when the temperature is below 27 °C (80 °F). Had the laboratory temperature been higher, he would never have obtained the two different crystals. Nevertheless, his was the sort of luck that happens to those prepared to take advantage of their luck. Most laboratories were and are cooler than 27 °C and many other people had looked at those same crystals before Pasteur. But he was the first one who looked carefully.

Pasteur's optically active solutions contained no crystals, only molecules, so their optical activity had to be due to the molecules themselves. Because the optical activity of the crystals depended on their shape, the activity of the molecules also had to depend on their three-dimensional shape. Two kinds of racemate crystals meant racemate molecules with two different shapes. Moreover, because the optical rotations were at exactly opposite angles, somehow or other some critical direction in the two different molecules was at exactly opposite angles. In other words, the molecules had to be mirror images of each other. But how? Try as he might, Pasteur could not explain how molecules could be mirror images of each other. Eventually, in 1860, he came up with the suggestion that the molecules involved were asymmetric, perhaps shaped like a helix, or with atoms at the points of an irregular tetrahedron. By that time, however, he had left stereochemistry to others. His attentions were now focused on bacteriology, where once again he was lucky.

Pasteur was not the only one to think of tetrahedral molecules. Alexander Michailovich Butlerow (1828–1886) in Russia proposed in 1862 that the carbon atom was a tetrahedron, with the valences directed to the center of the faces rather than to the corners, as we now visualize them. Oddly enough, Kekulé, who never accepted the reality of the atom, suggested in 1859 that carbon atoms were tetrahedral and in 1867 spoke of the valency units of carbon as being directed in space.

Stereoisomerism

The explanation of optical activity came suddenly in 1874 when two young men published the answer simultaneously. By coincidence they knew each other, and even worked for a time in the same laboratory, but kept their ideas about optical activity strictly to themselves and arrived at the same conclusion independently. Jacobus van't Hoff (1852–1911) was twenty-two years old at the time and Joseph-Achille Le Bel (1847–1930) was twenty-seven. Le Bel's work was more abstract, and van't Hoff's diagrams were more concrete and specific, although at first the world took no notice of either.

Van't Hoff suggested that the valences of a carbon atom are directed to the corners of a tetrahedron. A carbon atom with four different groups attached to it, that is, an asymmetric carbon, is therefore capable of existing in two different forms. He illustrated his idea with a great number of compounds, pointing out that derivatives of those with asymmetric carbons are also optically active. He explained racemic mixtures and meso compounds and, on the basis of asymmetric carbons, even explained cis–trans isomerism in compounds with double and triple bonds.

Van't Hoff sent copies of his work to Wurtz and others, but few chemists paid attention to it, most probably because they still doubted the existence of atoms and molecules. The deficiencies of the Dalton Atom theory and its inability to explain valence and electricity had led some to think in terms of the Boscovitch atom, which was one step away from no atom at all (*see* Chapter XV). In 1864 Lothar Meyer, who later codiscovered the Periodic System, was still skeptical about the atom, as was the great organic chemist August von Hofmann, even though molecular formulas and atomic weights were by now well established. In 1867 Kekulé wrote

> The question whether atoms exist or not has but little significance in a chemical point of view. Its discussion belongs rather to metaphysics. In chemistry, we have only to decide whether the assumption of atoms is an hypothesis adapted to the explanation of chemical phenomena.

In such an atmosphere, who would take seriously the work of two young men who not only considered that atoms were real but who talked about directions of valences and shapes of molecules?

In 1875, however, van't Hoff expanded his paper into a book and sent copies, with cardboard models of his tetrahedral carbon atoms, to a number of famous chemists, including Adolph von Baeyer, Butlerow, Hofmann, Kekulé, Frankland, Wurtz, Johannes Wislicenus, and Kolbe. This time, they sat up and took notice. Von Baeyer and Wislicenus were enthusiastic, but Berthelot and especially Kolbe rejected the idea in toto. The irascible Kolbe, who loved controversy, published a scathing, almost scurrilous, attack on van't Hoff and his work, proclaiming that it would be impossible to form a picture of the spatial arrangement of atoms. He called the idea "frivolous . . . phantasmagorical puffery . . . fantastic foolishness . . . fantasy . . . thoroughly unintelligible . . . shallow speculations . . . not far removed from belief in witches and ghost-rapping."

Kolbe's attack backfired, partly because it was so far-fetched and intemperate that it attracted the attention of chemists who might not otherwise have read van't Hoff's book. By 1894, Wislicenus in Leipzig could write that van't Hoff's views were almost universally accepted except by those who still did not accept the atomic hypothesis. Shortly thereafter, the revolution in physics demonstrated the real existence of atoms and subatomic particles.

References

1. Mauskopf, S. *Trans. Am. Phil. Soc.* **1976,** *66,* 12.
2. Lavoisier, A. *Acad. des Sciences* 1785 (1788) m 590 VII ii 656, cited by Partington, J. R., *The History of Chemistry;* Macmillan: London, New York, 1962; Vol. III, p 469.
3. Bertholet, C. *Obs. Phys.* **1785,** *XXVII,* 88; **1786,** *XXVIII,* 272; **1786,** *XXIX,* 389.
4. Berzelius, J. J. *Lärbok IV,* (3)(a), 1827, III i 151 (3)(d)VII; cited in Partington, J. R. op. cit., Vol. IV, p 252.
5. Gay-Lussac, J. L. *Ann. Chim.* **1814,** *XCI,* 149, cited by Partington, J. R., op. cit., p 256.
6. Geison, G.; Secord, J. *Isis* **1988,** *79,* 8.
7. Phillips, J. P. "Liebig and Kolbe, Critical Editors,"*Chymia* **1966,** *11,* 7 et seq.
8. Laurent, A. *Ann. Chim.* **1846,** *XVIII,* 266–298. Also Hunt, T. S. *Amer. J. Sci.* **1848,** *VI,* 173–178, cited by Partington, J. R. op. cit., p 422.
9. Williamson, A. "Papers on Etherification and the Constitution of Salts," *Alembic Club Reprint, 16;* University of Chicago Press: Chicago, 1906.
10. Frankland, E. *Phil. Trans.* **1852,** *CXLII,* 417.

11. Couper, A. S. "On a New Chemical Theory and Researches on Salicylic Acid," *Alembic Club Reprint*, University of Chicago Press: Chicago, 1933.
12. Kekulé, A. *Annalen* **1858**, *CLI*, 129, cited by Partington, J. R., op. cit., Vol. IV, p 556.

XV

The Divisible Atom

In spite of the successes in discovering complicated molecular arrangements, as late as the 1880s there was no knowledge of the interatomic forces holding molecules together, and no explanation of why atoms combined only in certain directions and certain numerical combinations. With the clarity of hindsight, we can say that nineteenth century chemists needed to know the internal structure of atoms in order to understand how they combine. Unfortunately, at the time, chemists were too involved with more specific problems to worry about the internal composition of atoms that might not exist, and if they did exist, were by definition indivisible.

Chemists pursued their own short-term goals, but studies in two areas eventually showed that there was some sort of internal atomic structure and that it was electrical in nature. These areas were electricity and periodicity, or similarity among properties of groups of atoms. Still, even among those who worked in these areas, few even speculated about internal atomic structure.

Back in the seventeenth century, Boyle had proposed that chemical atoms were made of prime matter and had different shapes. In the eighteenth century, Priestley had proposed atoms that "must be divisible and therefore have parts" (*1*). The polymath Roger Joseph Boscovitch (1711–1787), a Jesuit priest, philosopher, poet, linguist, mathematician, astronomer, and physicist, suggested that chemical atoms were composed of mathematical points surrounded by forces that were alternately attractive and repulsive. In developing this idea, he proved mathematically that Newton's impenetrable atoms could not exist, that in a collision between two impenetrable atoms, each would simultaneously be moving in two opposite directions—a manifest impossibility. Most chemists paid little attention, simply using the concept of indivisible, impenetrable chemical atoms as a tool.

Chemical Electricity

Galvani and Volta

Chemical investigations of electricity began as a consequence of what we would consider a biophysical discovery. In 1780 the Italian physician Luigi Galvani (1737–1798) had found that electricity caused muscle contractions[1]. He saw a frog muscle suspended on a metal hook twitch when lightning flashed. On investigating further, he found that, even in the absence of lightning, when a frog muscle and nerve were connected to each other by two wires, each of a different metal, the muscle twitched. The twitch was obviously the work of an electrical discharge. But what was the source of the discharge? Galvani, a physician, was primarily interested in anatomy and physiology, so he had no trouble deciding that somehow the electricity arose in the animal tissue. He called it "animal electricity."

Galvani's explanation was opposed by Count Alessandro Volta (1745–1827), a physicist. He rejected animal electricity, proposing instead that the source of the electricity was the junction of the two metals that made the electrical connection to Galvani's frog tissues. To demonstrate that, in the absence of animal material, electrical potentials could be produced by joining two different metals, he put a copper disk on top of a zinc disk. The copper now had a negative charge and the zinc a positive charge, and he observed a potential difference between them. (Our term for the phenomenon is *volta potential.*) Following up this observation, he made several such pairs of disks and placed one pair on top of another in a pile, with each pair of disks separated from the others by moistened paper or cardboard. When the disk on the top of the pile was connected by a wire to

[1]Galvani was a principled but unfortunate man. When the French under Bonaparte overran Italy, he refused to swear allegiance to the new regime. In consequence, he lost his position and was reduced to poverty.

the disk on the bottom of the pile, he observed a strong electrical discharge[2].

In 1800 Volta wrote of his discovery to the president of the Royal Society, Sir Joseph Banks. Banks, in turn, told two friends, William Nicholson (1757–1815) and Anthony Carlisle (1768–1840). Within a month, Nicholson and Carlisle had constructed a duplicate of Volta's "electrical pile," or battery, and discovered that when they connected a wire to the top disk and another to the bottom and put the connecting wires into water, they generated hydrogen and oxygen, in the ratio of two volumes of hydrogen to one of oxygen. This experiment was what Sir Humphry Davy called "the true origin of all that has been done in electrochemical science." It showed a direct connection between electricity and chemical action.

Davy and Faraday

Davy was deeply involved with electrochemistry from the start of his career. Using Volta's battery, he electrolyzed different molten salts and discovered six new elements: sodium, potassium, strontium, calcium, barium, and magnesium. Having used electricity to extract elements from their compounds, Davy naturally was disposed to think that chemical attraction between atoms is electrical in nature. He never really accepted the Dalton atom and, as time passed, his objections became stronger. In a paper in 1809 he wrote that he preferred the word "proportion" to "atom" or "atomic weight" (2). In 1814, with Michael Faraday, he analyzed diamond by combustion and concluded that it was pure carbon. Because graphite was also known to be pure carbon, this meant that there were two kinds of carbon, each with different physical properties. (This phenomenon is called "allotropy" and is quite common, as many elements being allotropic as are not.) Davy's discovery of the allotropy of carbon meant that graphite and diamond existed in two different crystal forms, even though this could not possibly be explained on the basis of Dalton's spherical atoms. By now, Davy was convinced that there were indeed many different kinds of atoms, other than Dalton's.

[2]As yet, electrical meters had not been invented. To test what we would now call a power supply, at least semiquantitatively, the observer connected two wires to the source of the electricity and then touched the ends to his tongue. If it really hurt, he had a good source of electricity. Scientific investigation at that time took a certain amount of physical stamina and intestinal fortitude.

The interplay between Davy and Michael Faraday (1791–1867) is fascinating. Faraday was a poor boy who worked as a bookbinder's apprentice. He attended evening lectures intended to uplift worthy members of the laboring classes, and with his employer's permission read books that came in for binding. By chance, someone gave him a ticket to Davy's lectures at the Royal Institute. Faraday was overwhelmed by both Davy and chemistry. He took copious lecture notes, wrote them up, bound them in a book, and sent it to Davy with a request for a job in chemistry. Davy interviewed him and hired him. Eventually Faraday far outshone Davy, which Davy on occasion resented.

Faraday's greatest achievements were in physics. He discovered electrical induction and the connection between magnetism and polarization of light (the Faraday effect). He was a founder of field theory. Actually, however, he was trained by Davy as a chemist and as a chemist, he discovered benzene, the "Kolbe hydrocarbon synthesis," codiscovered isomerism, and did much more. In collaboration with William Whewell, scientist, historian, and late in his career, the master of Trinity College, Cambridge, he coined the terms *ion, anode, cathode, anion, cation,* and *electrode.*

Here we are concerned with his discovery of the laws of electrochemical conduction and reaction, around 1834. These are (1) The amount of chemical decomposition (reaction) produced by an electrochemical current is directly proportional to the quantity of electricity passed, and only to that. (2) The amounts of different substances deposited or dissolved by passage of current are in direct proportion to their chemical equivalent weights. (3) One equivalent weight of any chemical substance is associated with a fixed amount of electricity (now known as the faraday).

Until Faraday's quantitative discoveries, electricity had been thought of as a fluid or an imponderable substance, but Faraday's laws established that there is a definite quantity of electricity associated with each atom, and implied strongly that electricity itself is in particle form. From our vantage point (but not his) Faraday had shown that neutral atoms consist of charged ions plus electrical particles.

At the time, chemists paid little attention to the implications of Faraday's electrochemical work. The Berzelius–Davy electrochemical theory of bonding was collapsing, and by 1840 few chemists were using the term *atom* without first qualifying it. Nevertheless, electrochemistry was too profitable a business to be neglected. Various batteries were developed, and in 1844 the first electroplating patents were issued. Electrolytic silver plate produced by electrolysis soon replaced the more expensive Sheffield plate produced by hammering sheets of silver onto heated copper articles.

Naturally, there was a good deal of scientific interest in the mechanism of conduction of current through electrolytic solutions. It was pretty well agreed that there had to be at least a few charged ions in solution to carry the current, but no agreement either on how they did it or even on what ions actually were. Mechanisms involving some sort of electrochemical "chains" were proposed, as were mechanisms in which the charges were produced by the action of the current and disappeared as soon as the circuit was open. No one seriously thought that large numbers of electrically charged particles could exist in solution in the absence of an applied voltage. After all, any positive and negative charges present would immediately combine with and neutralize each other. Moreover, because atoms were indivisible, the electric charges must somehow have come from electricity in the wires. Electrolytic conductance remained a puzzle.

Arrhenius

Clarification came a long time later, in 1884. Svante Arrhenius (1859–1927), a doctoral candidate at the University of Stockholm, wrote his thesis on the conductance of electrolytes (3). His experimental work was done in very dilute solutions, and, to explain his results, he proposed that electrolytic solutions contain large numbers of charged ions at all times, even in the absence of applied potentials. He agreed that oppositely charged ions attract and combine with each other, but he argued that the electrolyte molecules constantly dissociate, forming ions at the same rate as that at which the ions combine. Therefore, an equilibrium is established with ions always present. From his conductance measurements, Arrhenius calculated the degree to which the electrolyte dissociated and the equilibrium constant for the dissociation. He gave a kinetic derivation of the law of mass action, explained hydrolysis of salts, showed that the strength of an acid was proportional to the degree of dissociation, and even decided that the heat of neutralization of acids and bases was nothing but the heat of formation of water from hydrogen and hydroxyl ions. In 1887, Sir Oliver Lodge commented that Arrhenius's doctoral thesis was "really an attempt at an electrolyte theory of chemistry."

In 1903 Arrhenius received the Nobel Prize, mostly for the theory set forth in his thesis, but in 1884 his ideas were too daring. The chemical world was divided between the majority, who believed in Dalton's indivisible atoms, and an influential minority, who believed that atoms did not exist at all. Arrhenius's thesis implied that both sides were wrong, that atoms did exist, but were not indivisible. He immediately ran into a good deal of trouble in his thesis defense. First of all, his thesis examiners had difficulty believing that large numbers of electrically charged particles were floating around in solution even when there was no electrical current. Second, Arrhenius had made the tactical mistake of using sodium chloride as an example. This gave him the problem of explaining to his examiners why, although sodium chloride was stable, nontoxic, and actually an essential substance, solutions of sodium chloride contained the poisonous chlorine and the very reactive metal, sodium. Arrhenius patiently explained that the sodium and the chlorine were present not as neutral atoms but as charged ions, with very different properties. This explanation, however, produced another new problem for him. He now had to explain where the ionic charges came from. This, of course, he could not do, because any suggestion that a charge was transferred from one neutral atom to another meant that atoms were divisible. Chemists of the stature of Hofmann and Kekulé might believe that atoms were merely a convenient fiction, but it would not have been wise for the graduate student Arrhenius to declare that Dalton and Newton had been wrong. In any case, the examiners formed a low opinion of this revolutionary and highly speculative thesis. They passed him, but with such a low grade that he could not hope for an academic position.

Still, the thesis was published, and he sent copies to Lodge in England and to the great physical chemist Wilhelm Ostwald in Riga. At that time, Ostwald (1853–1932) was studying the rates of reactions that are catalyzed by acids (*3, 4*). He noticed Arrhenius's suggestion that highly conductive acid solutions contain a relatively large number of hydrogen ions. Ostwald reasoned that if this were true, these highly conductive acid solutions should be good catalysts for his reactions. So he tried more than thirty different acids as catalysts in the hydrolysis of methyl acetate and the inversion of sucrose, and he correlated the rates of the reactions with the conductivity of the acid used. In all cases, the rate of reaction was directly proportional to the conductivity of the acid. Arrhenius was not only qualitatively correct, but quantitatively correct.

Ostwald immediately went to Stockholm, offered Arrhenius both his scientific cooperation and a job, and used his influence to get him a teaching position at Upsala. Still, even after Ostwald's results on acid catalysis were published, not many people accepted the Arrhenius theory. There were simply too many unanswered questions[3]. Oddly enough, Ostwald, who worked so hard and so successfully for the ionic theory of electrolytes, at that time did not believe in atoms. Not until 1909 did he agree that atoms exist, becoming convinced by Jean Perrin's determination of the Avogadro number and by J. J. Thomson's definitive work on the electron.

Atomic Weights and Properties

Meanwhile, independent of electrochemistry, analytical and inorganic chemists had for years been looking at atomic weights and thinking about their significance. Possibly the first speculation about atomic weights was that of William Prout (1785–1850), an English physician. In 1815 he pointed out that most elements had atomic weights that were almost exact multiples of the atomic weight of hydrogen, the lightest atom. He suggested, therefore, that all elementary atoms might be made up of integral numbers of hydrogen atoms, somehow packed together.

Prout's hypothesis was such an attractive idea, pointing toward simplicity in natural law, that it was immediately taken up, and attempts were made to prove it. Thomas Thomson, the same man who had drawn attention to Dalton's work, suggested that where combining weights were not integral multiples of the weight of hydrogen, they might be in error. J. C. Galissard de Marignac (1817–1894) did a great deal of analytical work trying to show that the experimental atomic weights were wrong and that Prout was right, but eventually he gave up. Jean Servais Stas (1813–1891) was at first convinced that Prout was right but after years of careful work concluded that Prout was wrong, that the hypothesis was an illusion (5). Even so, in 1887, he suggested that it couldn't be pure chance that the weights were so close to integral multiples, that there must be something in back of it.

The first attempt at correlating atomic weights and properties came in 1829, shortly after Prout proposed his hypothesis. Johann Dobereiner (1780–1849) pointed out that there are groups of elements, which he called triads, that had similar chemical properties, and in which the atomic weight of the middle ele-

[3]Perhaps the last holdout was L. A. Kahlenberg, of the University of Wisconsin. As late as the 1940s, he was still publishing articles attacking the Arrhenius theory.

ment was just about the average of the other two. One such triad consisted of calcium, strontium, and barium, whose weights were 40, 88, and 137, respectively. Another triad was lithium, sodium, and potassium, whose weights were 7, 23, and 39, respectively. There were others. Dobereiner's relationships were treated as coincidences and not taken seriously, but the germ of what we now call group relationships was there.

In 1850 Max von Pettenkofer (1818–1901) observed another mathematical correlation. Groups of elements with similar properties had equivalent weights that differed from each other by multiples of 8 in Pettenkofer's correlation:

Li = 7	Mg = 12
Na = 23 = (7 + 2 x 8)	Ca = 20 = (12 + 1 x 8)
K = 39 = (7 + 4 x 8)	Sr = 44 = (12 + 4 x 8)
	Ba = 68 = (12 + 7 x 8)

A year later, in 1851, Dumas reported that there seemed to be several series of similar elements, again with orderly increments in atomic weight. For example, in Dumas's correlation:

P = 31
As = 75 = (31 + 44)
Sb = 119 = (31 + 2 x 44)
Bi = 207 = (31 +4 x 44)

Also in 1851, J. H. Gladstone (1827–1902) attempted to arrange the known elements in order of increasing atomic weights, but he found no patterns. In 1857, and again in 1864, William Odling (1829–1921) also attempted to find patterns in tables of elements arranged in order of increasing atomic weight.

John Newlands (1837–1898) in 1865 arranged a table of atoms in order of increasing equivalent weight and noted that there seemed to be a repetition with every eighth element. He proposed to call this the Law of Octaves. He noted that there were family resemblances, and even inverted the positions of tellurium and iodine, as Mendeleev later did.

The Periodic Table

The attempt to find order in the properties of the elements culminated in the Periodic Table of the elements, first published in 1869 by Dmitri Mendeleev (1834–1907). Lothar Meyer (1830–1895) arrived at the same conclusions, but didn't publish them until 1870. As had others, Mendeleev noticed that there seemed to be groups or families of elements with similar physical and chemical properties. He noticed also that in arranging atoms in order of increasing weight, there was a periodic repetition of properties. Unlike his precursors, however, he ignored numerical relationships, such as those of Newlands and of Dumas, and concentrated on devising a table that would graphically show both group and periodic relationships. He tried various arrangements, at first with the groups arranged horizontally and the weights vertically, but finally he settled on columns of atoms arranged from left to right in order of increasing weight, and from top to bottom in groups reflecting chemical and physical properties. When he had finished, it was immediately obvious that all of the thousands of physical and chemical properties of the elements and many of their compounds followed relatively simple qualitative and quantitative rules. Furthermore, from the position of the elements in the table, chemists could estimate their properties and even predict formulas of their binary compounds. The Periodic Table is probably the most useful summary of information in all chemistry.

Useful or not, Mendeleev's work was not immediately accepted. Chemists took little notice of his papers, or Meyer's. Mendeleev, however, had made some extremely daring predictions, several of which turned out to be true. (Some also turned out to be wrong, but these have been forgiven and forgotten.) In studying his table, he noted three places where elements seemed to be missing below boron. To him, this meant that these elements had not yet been found. He gave these undiscovered elements provisional names, and in 1871, from their positions in the Periodic Table, he predicted their physical and chemical properties, including valence, color, atomic weight, and density. In 1875 one of these elements, gallium, which he had named eka-aluminum, meaning below aluminum, was discovered. Its atomic weight and chemical and physical properties were quite close to the predicted values. This time chemists took notice. In 1879 scandium (eka-boron) and in 1886 germanium (eka-silicon) were discovered, each with almost the predicted properties. From then on, the Periodic System received total acceptance.

The Innermost Limits of Chemistry

Faraday had demonstrated that in an electrochemical reaction each atom was associated with a definite quantity of charge. Arrhenius had shown that in a conducting solution, atoms were very often electrically charged, again with a fixed quantity of charge. Presumably the charges were fixed amounts of electrical fluid, and perhaps they were contained within the neutral atom, but that was as much as could be guessed. The Periodic System showed clearly that the chemical and physical properties of the atoms seemed to follow a pattern, and that therefore there was some sort of internal atomic structure. But chemists could only speculate on what this structure depended. They had gone about as far as they could without the help of physicists.

Chemical reactions, as we know, involve primarily the outermost electrons and thus cannot reveal the interior structure of the atom. The tools needed for further progress—high-energy electric fields, X-rays, radioactive emanations, and spectra—were those of the physicists.

References

1. Priestley, J. *Disquisition on Matter and Spirit*; 1777; pp 5–6, cited in Partington, J. R., *The History of Chemistry*; Macmillan: London, New York, 1962; Vol. III, p 296.
2. Davy, H. *Phil. Trans.* **1810 C.**, *63*, 16; XI, 271, cited in Partington, J. R., op. cit., Vol. IV, p 160.
3. Palmaer, W. In "Svante Arrhenius," *Great Chemists*; Farber, E., Ed.; Interscience: New York, 1961; pp 1095–1109.
4. Ostwald, W. *J. Prakt. Chem.* **1884,** *XXX*, 93, cited in Partington, J. R., op. cit., pp 676–677.
5. Partington, J. R.; op. cit., pp 877–878.

Epilogue

THIS NARRATIVE COMES TO A HALT at about the year 1900. By then, in the hundred or so years after the death of Lavoisier, modern chemical science had been organized into the self-consistent intellectual framework it still retains. Chemists now knew that the chemical properties of compounds are determined by the arrangements of the various atoms and that chemical reactions are simply the shuffling and reshuffling of the various combinations of atoms.

The nineteenth century chemists had laboriously determined correct relative atomic and molecular weights and worked out formulas and three-dimensional structures of many molecules. All of the myriad chemicals had been found to be combinations of fewer than a hundred basic elements, and these, in turn, had been organized into Mendeleev's Periodic Table. Chemists could now infer the chemical and physical properties of each element from its location in the Periodic Table. In many cases they could guess the correct formulas and many of the simple properties of compounds from the location of their constituent atoms in the table. Each location had been characterized by an integral number, such as 2 or 5, and so the bewildering complexity of chemical substances was simplified and empirically related to a few integers. In 1900 no one yet knew why this was so.

By 1900 there were large-scale chemical industries. Enormous numbers of useful new substances had been discovered and were being produced with new methods and apparatus. Mass production of synthetic dyes brought beautifully colored fabrics within reach of the even the poorest. Paul Ehrlich was busy synthesizing chemicals to cure syphilis. Soon Fritz Haber would synthesize ammonia to make artificial fertilizers to feed people and would then turn his attention to making poison gas to kill people in World War I. Insecticides, plastics, anesthetics, munitions, and fuels were on the way, even though at that point, chemists had no clear understanding of the mechanisms involved in chemical reactions.

No one yet knew how covalent materials are bound together, although electrovalent materials were known to be held together by electrical charges. Chemists knew that the electrical charges come from individual atoms and that the atom, although indestructible, is nevertheless divisible into electrically charged particles. Arrhenius's inspired logic had been verified by the discovery of the electron. Soon the nuclear structure of the atom would emerge. Irving Langmuir and Gilbert N. Lewis would explain covalence on the basis of the electron-pair bond, justifying the theory postulated by Archibald Couper and August Kekulé. Erwin Schrödinger and Werner Heisenberg would soon invent wave mechanics, which would then rapidly be applied to the calculation of molecular shapes, energies, and rates of reaction.

As these lines are being written, it is almost the end of the twentieth century. Computers and microprocessers are utterly transforming the study of chemistry. College sophomores perform mathematical operations beyond the powers of the greatest of the nineteenth century chemists. Technicians routinely run procedures that would have been impossible in 1900.

Research emphasis has shifted from the shapes of small molecules to the shapes of tiny subatomic regions in which electrons are most probably located and to the shapes of huge molecules containing hundreds of thousands of atoms. Verbal descriptions are giving way to mathematical equations, and the more closely the results of the equations correspond to experimental fact the further the equations seem to depart from what used to be common sense. In introductory college courses, students begin the study of chemistry not with the description of chemicals and reactions but by learning Schrödinger's differential equations, the shapes of atomic and molecular orbitals, and the distribution of electrons in the various atomic subshells.

Yet the science of chemistry is still the same and its objectives remain the same. Today's chemists still read and understand the work of their nineteenth century precursors. We still move in the directions they pointed out. But there is a difference between us and them. They created the intellectual framework in which we operate. They created the viewpoints we learn at the very beginning of our studies. When we look at a substance and think of its formula or look at a solution and think of ions and molecules rotating, stretching, twisting, swirling around, colliding, and separating, it is because they have taught us to do so. Their concepts, a century or two later, are still valid.

The chemistry of the future is anybody's guess, but of one thing we can be sure. The chemists of the future will solve their problems with the same mental equipment as the great chemists of the past.

Suggestions for Additional Reading

Historical Background

Bautier, R. H. *Economic Development of Medieval Europe*; Harcourt Brace, Jovanovich: New York, 1971.

Bernstein, J. *Experiencing Science*; Basic Books: New York, 1978.

Bowra, M. *Classical Greece*; Time-Life Books: New York, 1965.

Braudel, F. *The Structures of Everyday Life*; Harper and Row: New York, 1981.

Butterfield, H. *The Origins of Modern Science*; Free Press, Macmillan: New York, 1966.

Campbell-Watson, R. *Arabian Medicine and Its Influence on the Middle Ages*; AMS Press: New York, 1973. Reprinted from the 1926 edition.

Casson, L. *Ancient Egypt*; Time-Life Books: New York, 1965.

Clagett, M. *Greek Science in Antiquity*; Collier-Macmillan: New York, 1969.

Conant, J. B.; Nash, L. K.; Roller, D. H. *Harvard Case Histories in Experimental Science*; Harvard University Press: Cambridge, MA, 1950–1954.

Crombie, H. R. *Medieval and Early Modern Science*; Doubleday Anchor: New York, 1959.

Dampier, W. C. *A Shorter History of Science*; Meridian Books: New York, 1957.

Daumas, M. *A History of Technology and Invention*; Crown: New York, 1969.

Farrington, M. *Greek Science*; Penguin Books: Harmondsworth, Middlesex, England, 1947.

Forbes, R. J. *Studies in Ancient Technology*; Brill: Leiden, Netherlands, 1965. *Metallurgy in Antiquity*; Brill: Leiden, Netherlands, 1950. Forbes, R. J.; Dijksterhuis, E. J. *History of Science and Technology*; Penguin Books: Baltimore, MD, 1967.

Garraty, J. A.; Gay, P. *Columbia History of the World*; Harper and Row: New York, 1972.

Gillispie, C. C. *The Edge of Objectivity*; Princeton University Press: Princeton, NJ, 1960.

Hall, A. R. *The Scientific Revolution*; Beacon Press: Boston, MA, 1954.

Hawkes, J. *The First Great Civilizations*; Random House: New York, 1973.

Hawthorne, J. G.; Smith, C. S. Translation of *Theophilus on Divers Arts*; Dover Publications: New York, 1973. Reprint of edition published by University of Chicago Press: Chicago, 1963.

Heer, F. *The Medieval World*; Mentor Books: New York, 1962.

Herm, G. *The Phoenicians*; William Morrow: New York, 1975.

Hoover, H. C.; Hoover, L. C. Translation of *De Re Metallica*; by Georgius Agricola, Dover Publications: New York, 1950. Reprint of first edition published in *Mining Magazine*, London England, 1912.

Jung, C. J. "Psychology and Alchemy," Part III, *Religious Ideas in Alchemy*; Parthenon Books: New York, 1953.

Kline, M. *Mathematical Thought from Ancient to Modern Times*; Oxford University Press: New York, London, 1972.

Kramer, S. N. *History Begins at Sumer*; Thames and Hudson: London, England, 1958. *The Sumerians, Their History, Culture and Character*, University of Chicago Press: Chicago, 1963. *Cradle of Civilization*, Time-Life Books: New York, 1967.

Kuhn, T. S. *The Structure of Scientific Revolutions*; University of Chicago Press: Chicago, 1970.

Knauth, P. *The Ancient Smiths*; Time-Life Books: New York, 1974.

Latouche, R. *The Birth of Western Economy*; Methuen: London, England, 1961.

Levey, M. *The Medical Formulary of al-Kindi*; University of Wisconsin Press: Madison, WI, 1966. *Chemistry and Chemical Technology in Ancient Mesopotamia*, Elsevier: Amsterdam, Netherlands, 1955. *Al-Samarquandi*, University of Pennsylvania Press: Philadelphia, 1967.

Mason, S. F. *A History of the Sciences*; Collier-Macmillan: New York, 1961.

Muller, F. *Freedom in the Ancient World*; Harper and Row: New York, 1961.

Needham, J. *Clerks and Craftsmen in China and the West*; Cambridge University Press: Cambridge, England, 1970. *Science and Civilization in China*, Cambridge University Press: Cambridge, England, 1954-1974.

Partington, J. R. *History of the Greek Fire*; Heffer: Cambridge, England, 1960.

Rostovzeff, M. *Economic and Social History of the Hellenistic World*; Clarendon Press: Oxford, England, 1941.

Singer, S. et al *A History of Technology*; Oxford University Press: New York, London, 1954–1958. *A Short History of Scientific Ideas to 1900*, Oxford University Press: New York, London, 1959.

Trevor-Roper, H. *The European Witch Craze of the 16th and 17th Centuries*; Harper and Row: New York, 1956.

Underwood, E. A. *Science, Medicine and History*; Oxford University Press: New York, 1975.

White, L. T. *Medieval Technology and Social Change*; Oxford University Press: New York, London, Oxford, 1964.

Chemical Background

Alembic Club Reprints 1–20; The Alembic Club: Edinburgh, Scotland.

Ambix, Journal of the Society for the History of Alchemy and Chemistry, 1937 to present.

Benfy, O. T. *Classics in the Theory of Chemical Combination*; Dover: New York, 1963.

Clow, A.; Clow, N. *The Chemical Revolution*; Batchworth Press: London, England, 1952.

Crosland, M. A. *Historical Studies in the Language of Chemistry*; Dover: New York, 1978. Reprint of edition published by Heinemann Educational Books: London, England, 1962.

Debus, A. G. *The English Paracelsians*; Franklin Watts: New York, 1965. *The Chemical Philosophy*, Science Library Publications: New York, 1977.

Dobbs, B. J. T. *The Foundations of Newton's Alchemy:* or "The Hunting of the Greene Lyon"; Cambridge University Press: Cambridge and New York, 1975.

Eliade, M. *The Forge and the Crucible: The Origins and Structure of Alchemy*; Harper and Row: New York, 1962.

Farber, E. *Great Chemists*; Interscience: New York, 1961. *The Evolution of Chemistry*, Ronald Press: New York, 1952.

Freund, I. *The Study of Chemical Composition*; Cambridge University Press: Cambridge, England, 1904.

Gillispie, C. C. *Dictionary of Scientific Biography*; Charles Scribner's Sons: New York, 1970–1980.

Guerlac, H. *Lavoisier, the Critical Year*; Cornell University Press: Ithaca, NY, 1961.

Hall, M. B. *Robert Boyle and Seventeenth Century Chemistry*; Cambridge University Press: Cambridge, England, 1958.

Hannaway, O. *The Chemists and the Word: The Didactic Origins of Chemistry*; Johns Hopkins University Press: Baltimore, MD, 1975.

Holmyard, E. J. *Alchemy*; Penguin Books: Harmondsworth, Middlesex, England, 1957. *The Makers of Chemistry*, Clarendon Press: Oxford, England, 1945.

Ihde, A. *The Development of Modern Chemistry*; Harper and Row: New York, 1964.

Jaffe, B. *Crucibles, The Story of Chemistry*; Fawcett World Library: New York, 1967.

Lavoisier, A. *Elements of Chemistry*; Dover: New York, 1965. Unabridged reproduction of 1790 translation by Robert Kerr.

Leicester, H. M. *Historical Background to Chemistry*; Dover: New York, 1971. Reprint of edition of John Wiley and Sons: New York, 1956. *Chymia: Annual Studies in the History of Chemistry*, University of Pennsylania Press: Philadelphia, 1948–1967.

Lockemann, G. *The Story of Chemistry*; Philosophical Library: New York, 1959.

McKie, D. *Antoine Lavoisier, Father of Modern Chemistry*; Lippincott: Philadelphia, 1936.

Morselli, M. *Amedeo Avogadro*; D. Reidel: Dordrecht, Boston, Lancaster, 1984.

Multhauf, R. P. *The Origins of Chemistry*; Franklin Watts: New York, 1967.

Needham, J.; Gwai-Djen, L. *Science and Civilization in China; Volume V. Chemistry and Chemical Technology*; Cambridge University Press: Cambridge, England, 1974.

Pagel, W. *Paracelsus*; Karger: Basel, Switzerland, 1958.

Partington, J. R. *History of Chemistry*; St. Martin's Press: New York, 1961–1970. *Short History of Chemistry*, Macmillan: London, New York, 1939.

Read, J. *Through Alchemy to Chemistry*; G. Bell and Sons: London, England, 1957.

Rousseau, G. S.; Porter, R. *The Ferment of Knowledge*; Cambridge University Press: Cambridge, England, 1980.

Russell, C. A. *Recent Developments in the History of Chemistry*; Royal Society of Chemistry: London, England, 1985.

Singer, C. S. *A History of Technology*; Oxford University Press: New York, London, 1957. *The Earliest Chemical Relations of Economics and Technology Illustrated from the Alum Trade*; The Folio Society: London, England, 1958.

Stillman, J. M. *The Story of Alchemy and Early Chemistry*; Dover: New York, 1960, reprint of 1924 edition, Appleton: New York.

Szabadvary, F. *History of Analytical Chemistry*; Pergamon Press: Oxford, England, 1966.

Taylor, F. S. *The Alchemists, Founders of Modern Chemistry*; H. Schuman: New York, 1949. *A History of Industrial Chemistry*; H. Schuman: New York, 1957.

Thackray, A. O. *John Dalton, Critical Assessment of His Life and Science*; Harvard University Press: Cambridge, MA, 1972.

Ware, J. R. *Alchemy, Medicine and Religion in the China of A.D. 320*; The Ne-i P'ien of Ko Hung, Dover: New York, 1981. Reprint edition of MIT Press: Cambridge, MA, 1966.

Wolf, H. *The Analytic Spirit: Essays in the History of Science*; Cornell University Press: Ithaca, NY, 1981.

Index

A

C

E

G

H

I

J

M

N

O

P

Q

R

T

U

V

Production: Donna Lucas and Beth Pratt-Dewey
Indexing: Janet S. Dodd
Acquisition: Robin Giroux

Printed and bound by Maple Press, York, PA

DATE			